The Ramsar Convention on the Conservation of Wetlands

A Legal Analysis of the Adoption and Implementation of the Convention in Denmark

The Ramsar Convention on the Conservation of Wetlands

A Legal Analysis of the Adoption and Implementation of the Convention in Denmark

Veit Koester

Ramsar Convention Bureau
International Union for Conservation of Nature and Natural Resources
1989

First published in Danish as "Ramsar-konventionen om beskyttelse af vådområder — en retlig analyse af konventionens indgåelse og opfyldelse" (ISBN no.: 87-503-7403-6) in August 1988 by the Danish National Forest and Nature Agency under the Danish Ministry of the Environment and subsequently amended and updated.

An abbreviated version was published in the Danish magazine for the legal profession, "Juristen", 1989, pp. 8–32.

Translation: IFF Co., Copenhagen
The translation for the English edition has been financed by the Danish Forest and Nature Agency, and edited by Malcolm Forster.

Photos: Poul Hald-Mortensen, pages: IX, 39, 96
Flemming Pagh Jensen, pages: 2, 42
Palle Uhd Jepsen, pages: 83, 105
Jan Kofod Winther, pages: XII, 20, 44

ISBN No.: 2-88032-999-X

Printed by: Daemisch-Mohr GmbH, Siegburg

The Ramsar Convention

The Convention on Wetlands of International Importance especially as Waterfowl Habitat, sometimes also known as the Ramsar Convention from its place of adoption in 1971 in Iran, is an intergovernmental treaty which provides the framework for international cooperation for the conservation of wetland habitats.

Because wetlands are very important for ecological processes as well as for their rich flora and fauna, the broad objectives of the Convention are to stem the loss of wetlands and to ensure their conservation. To meet these objectives the Convention places general obligations on Contracting Parties relating to the conservation of wetlands throughout their territory and special obligations pertaining to those wetlands which have been designated in a "List of Wetlands of International Importance".

The Convention entered into force in late 1975 following the accession of the seventh Party, Greece. It now has Contracting Parties from all regions throughout the world.

The United Nations educational, scientific and cultural organization (Unesco) serves as depositary for the Convention. The secretariat, or Bureau, is an independent body administered by the International Union for Conservation of Nature and Natural Resources (IUCN) and the International Waterfowl and Wetlands Research Bureau (IWRB). Its headquarters are located in Gland, Switzerland, with a branch at Slimbridge in the United Kingdom.

International Union for Conservation of Nature and Natural Resources

Founded in 1948, IUCN – The World Conservation Union – is a membership organisation comprising governments, non-governmental organisations (NGOs), research institutions, and conservation agencies in 120 countries. The Union's objective is to promote and encourage the protection and sustainable utilization of living resources.

Several thousand scientists and experts from all continents form part of a network supporting the work of its six Commissions: threatened species, protected areas, ecology, sustainable development, environmental law, and environmental education and training. Its thematic programmes include tropical forests, wetlands, marine ecosystems, plants, the Sahel, Antarctica, population and sustainable development, and women in conservation. These activities enable IUCN and its members to develop sound policies and programmes for the conservation of biological diversity and sustainable development of natural resources.

About the Author

Veit Koester is a leading international environmental lawyer from Denmark. A graduate of the legal faculty of the University of Copenhagen, Mr. Koester serves as the Head of the Ecological Division of the National Forest and Nature Agency of the Danish Ministry of the Environment. He has been one of the main authors of the present Danish nature protection legislation and has been chiefly responsible for work on the revision of the Danish Nature Conservation Act. His office has day-to-day authority, i.a. for administering the provisions of the Conservation of Nature Act dealing with general protective measures for the biotopes in Denmark and for Danish implementation of global and European nature conservation conventions and Directives.

Mr. Koester has had considerable [illegible] international [illegible] conservation [illegible] Convention [illegible] He has [illegible] served as Head of the Danish delegation at several international conferences including meetings of the Conference of the Contracting Parties to the Ramsar Convention, the [illegible] Migratory Species and CITES Convention and has been elected as Chairman of [illegible] international meetings.

[illegible] International Union for Conservation of Nature and Natural Resources (IUCN) and is an expert member of the IUCN Commission on Environmental Policy, Law and Administration (CEPLA) and the International Council of Environmental Law (ICEL). He is presently also the President of the [illegible] Standing Committee of the Bern Convention.

Mr. Koester has authored several books and articles in the field of environmental law at both national and international level. Among these works has been the 1980 publication for IUCN of Nordic Countries Legislation on the Environment [illegible]

About the Author

Veit Koester is a leading international environmental lawyer from Denmark. A graduate of the legal faculty of the University of Copenhagen, Mr Koester serves as the Head of the Ecological Division of the National Forest and Nature Agency of the Danish Ministry of the Environment. He has been one of the main authors of the present Danish nature protection legislation and has been chiefly responsible for work on the revision of the Danish Nature Conservation Act. His office has day-to-day authority, i. a. for administering the provisions of the Conservation of Nature Act dealing with general protective measures for the biotopes in Denmark and for Danish implementation of global and European nature conservation conventions and Directives.

Mr Koester has had considerable personal involvement in the drafting and the conclusion of several international nature conservation instruments including the World Heritage Convention, the Migratory Species Convention, the European Convention (Bern), and the European Community Birds Directive. He has served as head of the Danish delegation to several international conferences including meetings of the Conference of the Contracting Parties to the Ramsar Convention, Migratory Species and CITES Convention and has been elected as Chairman or Vice-Chairman at several of these international meetings.

Mr Koester is a member of the Council and Bureau of the International Union for Conservation of Nature and Natural Resources (IUCN), and is an expert member of the IUCN Commission on Environmental Policy, Law and Administration (CEPLA) and the International Council of Environmental Law (ICEL). He is presently also the President of the Standing Committee of the Bern Convention.

Mr Koester has authored several books and articles in the field of environmental law at both national and international level. Among these works has been his 1980 publication for IUCN on "Nordic Countries Legislation on the Environment".

Table of Contents

he Black-headed Gull (Larus ridibundus) is one of the wetlands' most common nd adaptable species. It can be found in urban parks, small bogs and lakes, as ell as near fiords and the open sea.

Preface

The Wetlands, or "Ramsar", Convention is the oldest of the global nature conservation treaties, and the only one to deal with a particular ecosystem type.

The Convention covers a very wide variety of wetland habitats including rivers, lakes, ponds, marshes, coastal areas, estuaries, bogs and even coral reefs. The Convention places numerous obligations upon the Contracting Parties for the wise use of wetland habitats, for special conservation requirements for wetland sites designated onto the "List of Wetlands of International Importance", for the creation of wetland reserves, for international cooperation, for shared waterbodies and shared wetland species, for the wardening of sites to enhance waterfowl numbers, for the training of wetland personnel and for the promotion of general public awareness for conservation.

Denmark was one of the first Contracting Parties to the Wetlands Convention and has had considerable experience in implementing its provisions. This experience is particularly interesting in view of Denmark's sophisticated land use planning system. Although not directly applicable elsewhere, the Danish experience should provide valuable insights to other Contracting Parties and to other States preparing to join the Convention.

Veit Koester, an international environmental lawyer, is eminently well qualified to make this analysis. He has served for several years as the officer at national level responsible for Danish implementation of the Convention and has been elected as Chairman and as Vice-Chairman at global Ramsar meetings of the Conference of the Contracting Parties.

We are indebted to the Danish Ministry of the Environment for supporting the production of this volume and to CEPLA member Professor Malcolm Forster of Southampton University for his extensive assistance in the editing of the English translation of this work.

We are delighted to present Mr. Koester's study as a joint Ramsar Convention/IUCN publication and extend to him our deep appreciation for this work.

W.E. Burhenne
Chairman
IUCN Commission on Environmental Policy, Law and Administration

D. Navid
Secretary General
Ramsar Convention Bureau

Salt meadows with tidal channels on the Skallingen peninsula in the Wadden Sea which in 1987 became the 27th Danish site on the Ramsar Convention's List of Wetlands of International Importance.

. Introduction

.1. The Ramsar Convention

he Convention of February 2, 1971 on Wetlands of International Importance specially as Waterfowl Habitat (the Ramsar Convention), entered into force on ecember 21, 1975. Denmark ratified the Convention on December 19, 1977 — ee the Danish Ministry of Foreign Affairs' Executive Order No. 26 of April 4, 978.

his was probably the first time that Denmark undertook an international obliga- on to preserve certain large areas of the country, an obligation which has to ome extent restricted national sovereignty over land use. Application of the onvention in Denmark thus involves aspects of international, constitutional and dministrative law. The purpose of this analysis is to examine these aspects in ore detail, 10 years after Danish ratification. This is particularly important ecause, as will be seen, the Convention has played, and continues to play, a ajor role in the determination of regulations, land-use planning, and administra- e practice, and of course in particular in the conservation of the 27 designated anish Ramsar sites (Chapter 4.2.). Nonetheless, the Convention has attracted ry little attention in legal and administrative literature (1).

s well as examining the relevant regulations in the broadest sense, this review ill also focus on planning and administrative practice in individual cases, thus abling a fairly accurate impression to be gained of the current conservation atus of Danish Ramsar sites, at least as to immediate or direct physical interfer- ce (as opposed to pollution). In order to avoid burdening the text unnecessarily, number of special problems, such as detailed reviews of actual decisions and her special circumstances, have been relegated to the notes, which therefore ntain not only references to sources but in many respects supplement or nplify the text itself.*

2. The EC Directive on the Conservation of Wild Birds

will appear from Chapter 3.4., all the Danish Ramsar sites are included in the 1 EC Bird Protection Areas which have been designated in Denmark in pur- ance of the EC Directive on the Conservation of Wild Birds. The legal signifi- nce of this coincidence will not be considered further, since the implementation the Directive in domestic law, and its administration, lie beyond the scope of s analysis. Nonetheless, the EC Bird Protection Areas will be touched upon to a rtain extent. Furthermore, conservation obligations under the Directive are no s extensive than those applicable to designated sites under the Ramsar Con-

---

or practical reasons the index to the book also comprises the notes. Please note also that the umeration of the footnotes follows the order of the original Danish text. Furthermore, a number of otes have been added as an explanation of Danish legislation and to update the original Danish text, hich thus for the most part has been updated to approx. mid-1989.

vention. There is therefore no doubt that the conclusions as to the obligations existing in respect of Ramsar sites, as reflected in regulations and administrative practice, are in general equally valid for those almost 20 pct. of EC Bird Protection Areas which are not also Ramsar sites (26). The extent to which the conclusions concerning international and constitutional law may validly be applied to those *sites* is quite another question, which will not be pursued further in this context.

The Tufted Duck (Aythya fuligula) is present in large numbers in East Danish lakes and coastal areas. In total 100,000–200,000 birds winter here. The breeding population is around 500–650 pairs. Five Danish Ramsar sites are of international importance for this species.

2. The Ramsar Convention

2.1. Background

The importance of wetlands to society, both in their intrinsic value and as an indispensable element of the ecological cycle (2) has long been recognized by biologists and other natural scientists. Nonetheless, international efforts to counter the rigorous destruction of wetlands (4) did not seriously commence (5) until the early 1960s, when draining and land reclamation were at their peak in Denmark, stimulated by large government subsidies (3). A number of international conferences and meetings of technical experts resulted in the drafting of the Ramsar Convention, named after the Iranian town in which the Convention was signed on February 2, 1971 (6).

2.2. Purpose and Jurisdiction

The Ramsar Convention is unique in that it remains the only global convention the objective of which is to protect and conserve a particular type of ecosystem and the flora and fauna (especially waterfowl) dependent upon it.

Wetlands are defined in Article 1 of the Convention as areas of marsh, fen, peatland or water, whether natural or artificial, permanent or temporary, with water that is static or flowing, fresh, brackish or salt, including areas of marine water the depth of which at low tide does not exceed six metres. According to Article 2(1) riparian and coastal zones adjacent to wetlands may be incorporated in areas designated for inclusion in the International List — see Chapter 2.3. The same applies to bodies of marine water deeper than six metres at low tide, especially where they are important as waterfowl habitats.

Although (as the title of the Convention suggests) waterfowl form a very important consideration, it is essential to bear in mind that the Convention envisages the conservation of all of flora and fauna dependent upon a particular wetland. This aspect is often forgotten in practice, although it is clearly stated for example in the preamble to the Convention and a number of its Articles (cf. e.g. Articles 2(2); 4(3); 5(2) and 6(2) (d)).

2.3. Key Obligations

The most important obligations are those concerning land-use planning, the designation of one or more wetlands for inclusion in the "List of Wetlands of International Importance" and their conservation, and finally the promotion of the protection of wetlands in general.

In the sphere of physical planning the Convention requires that planning must be carried out so as specifically to promote the conservation of the wetlands included in the List, and generally to promote as far as possible the wise use of all wetlands (Article 3(1)). An interpretation of this provision adopted at the First Conference of the Parties amplified this obligation to a requirement to promote the preservation of the ecological character of wetlands, i.e. by preventing their destruction, change or pollution (7).

Article 2(4) requires that each Contracting Party, when acceding to the Convention, shall designate at least one wetland for inclusion in the List of Wetlands of International Importance. The criteria for these sites are set out in Article 2, as amplified and supplemented at meetings and conferences of the parties (8). In cases of "urgent national interests", it is possible to delete wetlands from the list or restrict their boundaries (Article 2(5)), as the inclusion of a wetland in the List does not prejudice the exclusive sovereign rights of the states (Article 2(3)). Such deletions or restrictions must be notified to the Bureau of the Convention at the earliest possible time (Article 2(5)). Furthermore, the Contracting Party concerned is obliged to endeavour to compensate for the loss as far as possible (Article 4(2)). It is worth noting that no site has yet been deleted from the List (9), although boundary reduction has been discussed in Denmark on at least one occasion — see Chapter 5.6.5.2.

The last key obligation is the promotion of the conservation of wetlands by establishing nature reserves and providing adequately for their wardening. This obligation applies whether or not the sites are included in the List (9a).

The question of whether a certain activity conflicts with these conservation obligations depends primarily on an assessment of the impacts of the activity. The central question is whether the site can continue to provide a habitat for its existing flora and fauna (and primarily for waterfowl) to the same extent as before. In substance, this requires an ecological evaluation of the impacts of the activity, especially any detrimental effects. The obligation to prevent changes will presumably lead to the prohibition of activities where there is a risk of negative impacts, even though there may be uncertainty as to the precise nature of those consequences. This at any rate is how the Convention has been understood by the Danish Government in relation to sites designated under Article 2(4) (73). However, in Danish administration practice evaluations on detrimental effects on the landscape itself (scenic values) have also played a certain role in the decision-making process. The form and extent of the ecological and other technical evaluations, which are a fundamental element of the implementation of the Convention, do not fall within the bounds of this study.

2.4. The Nature of the Obligations

From a strict legal viewpoint, it must be acknowledged that the obligations are rather vague ("to promote ... as far as possible the wise use of wetlands in their territory" — Article 3(1); "promote the conservation" in Article 4(1); "it should as far as possible compensate for any loss of wetland resources" in Article 4(2)). Nonetheless, the impact of the Convention has been positive. This is probably due to the readiness of a number of Contracting Parties to accept (on the whole) the obligations set out in the Convention while, in cases of encroachment on wetlands, it has been possible to use the Convention as a national and international platform for public opinion, naturally involving principally the "green" organizations (10). Furthermore, the requirements of the Convention are supplemented by general principles of international law, "soft law" and other legal instruments — Chapter 3.

In this context it is also important to emphasize that the Ramsar Convention generally does not specify a particular method for its implementation (with the partial exception of the obligation in Article 2(4) to designate at least one site to the List of Wetlands of International Importance): results are more important than means. As a necessary minimum, it is sufficient that the states in their application of the law respect the international law obligations which are contained in the Convention (10a).

2.5. Amendments to the Convention

The Ramsar Convention was the first global nature conservation convention. For this reason there was no special experience on which to draw when the Convention was drafted. As a result the Convention suffered from the outset from certain practical inadequacies. It has taken a long time to remedy these, and their effects will probably never be completely eradicated.

By the early 1980s, it was already clear that the provisions concerning the Convention's secretariat were far too weak and that this problem could not be solved without establishing a budget and a system for the partial financing thereof (11). The Convention as drafted did not, however, contain an amendment procedure. It was therefore necessary first to draw up (1980–82) and adopt an amending protocol at an extraordinary Conference of the Parties on December 3, 1982 (12). The required number of accessions (including that of Denmark) to enable the protocol to come into force were not received until October 31, 1986.

Meanwhile, at the Second Ordinary Conference of the Parties in 1984, work had begun on preparing the substantial amendments to the text of the Convention so that, in accordance with the amendment procedure in the protocol, amendments could be put forward for negotiation and possible adoption at the Third Conference of the Parties, which took place in the early summer of 1987. In the event, these amendments were successfully adopted after a number of legal and technical difficulties, such as the fact that by no means all the Parties to the Convention had acceded to the protocol.

It may take several years before a sufficient number of Parties accede to the amendments to the Convention to enable them to come into force. On the othe hand, however, agreement has been reached by consensus on the immediate implementation of the most important of the principles contained in the amendments from now on (13).

2.6. Status

On January 1, 1989 more than 50 countries had acceded to the Ramsar Convention, the majority having also acceded to the amendment protocol.

Member States include all the Nordic and most other Western European countries (including all the EC countries, with the exception of Luxembourg) and severa East European states (including the USSR). Major non-European states include Canada, the USA, Australia, Japan and India. There is considerable underrepresentation of developing countries, and of countries in the Southern Hemisphere At the Third Conference of the Parties (in 1987) the search for a solution to this problem was an important topic.

On June 1, 1988 rather more than 400 wetlands with a total area of approx. 285,000 km^2 had been included in the International List (13a). The designated sites vary considerably in size, from e.g. Switzerland with 2 sites of a total of almost 20 km^2, to Canada with 28 sites of a total of almost 130,000 km^2. In terms of area, even when the sites on Greenland are excluded, Denmark ranks sevent (and highest among the European states, with the exception of the USSR) with 27 sites of a total of a little over 7,400 km^2 (of which the Wadden Sea and adjacent coastal areas account for a little over 1,400 km^2 — see also Chapters 3.1.; 4.2. and 6.1. and notes 21 and 32). This emphasizes the great importance of the country for migratory waterfowl in particular. In addition there are 11 sites in Greenland of a total of approx. 10,000 km^2 (of which the greater part are land areas) designated by the Greenland Home Rule and notified to the List of Wetlands of International Importance in the spring of 1987 (1, B).

3. Standards of International Law, Soft Law, Other Conventions, etc.

3.1. Principles of International Law

As indicated in Chapter 2.3., the extremely vague obligations in the Ramsar Convention cannot be evaluated and interpreted in isolation, since they have to be considered together with other principles of customary international law, as well as any so-called "soft law" evolving subsequent to the adoption of the Convention. It is important to remember this when analysing how the Convention has been implemented in Denmark — see below.

Such principles of international law must be presumed to include some of the fundamental provisions of the 1972 Stockholm Declaration from the UN Conference on the Human Environment (14), such as Principle 2 on the conservation of representative natural ecosystems and Principle 4 on the protection and wise management of endangered habitats, amongst other things.

Principles of a similar kind, although in more detailed and practical form, are to be found in the World Charter for Nature (15) adopted by the UN on October 29, 1982 by 111 votes to 1 (the USA), which together with the Stockholm Declaration and other similar texts has also been described as providing the elements of an International Constitution for the World Environment (16).

A further fundamental international document is the UN Convention on the Law of the Sea which has been signed by a very large number of states and can therefore in many respects justifiably be said to reflect generally acknowledged international law principles (17), even though it has not come into force. A relevant example of these principles is Article 194(5), Articles 56(1)(b) (iii) and 192 (18) of the Convention which imply an obligation to protect and conserve rare or vulnerable ecosystems (19).

These considerations are of particular emphasis in respect of "transboundary" nature areas, since the principle that a state must ensure that its activities are not detrimental to areas situated in other countries or outside its national jurisdiction is assumed to be an element of customary international law (20). Positive and constructive application of this principle forms the basis for Danish/Dutch/German cooperation on the protection of the Wadden Sea, a very important wetland, not only to these countries but also in international terms (21). Denmark and Holland have included this wetland in the List of Wetlands of International Importance, while it is expected that West Germany will follow suit.

Similarly, a factor which can also serve to make control more stringent, concerning wetlands in particular, is the "international responsibilities" (Article 2(6) of the Ramsar Convention) as to migratory species, such as migratory birds, which at different stages of their lives or annual cycles are dependent on such areas.

These "responsibilities" are also reflected in other international instruments, such as e.g. the Convention on the Law of the Sea (Articles 64; 65, 120 and 118) and the Bonn Convention on the Conservation of Migratory Species of Wild Animals (22).

Due to its geographical location and nature resources Denmark is of very great importance to a number of migratory species of waterfowl, including rare endangered or vulnerable species, and this factor is therefore of very particular importance.

Moreover, there is no clear delineation between an axiom of customary international law on the protection of our common heritage and the recognized principle in customary international law "of equitable utilization of the environment", which means that states must use the environment in such a manner that other states may equally enjoy its use (22a).

3.2. Soft Law

In an international context, the term "soft law" is applied to declarations, recommendations, resolutions, etc., which are not binding in a legal sense but which nonetheless, depending on the forum in which they have been adopted, can often have a certain moral or political force. Examples include resolutions and similar adopted by the Council of Europe, the OECD Council or Conferences of European Ministers for the Environment. Several such instruments are concerned particularly with the conservation of wetlands, or with special planning considerations related thereto (23).

Recommendations and resolutions adopted at conferences of the parties to the Convention (cf. Chapter 2.3. above) may be included in this group. These instruments, however, may tend gradually to achieve a binding effect (7).

3.3. Other Conventions

Where states are parties to both the Ramsar Convention and to other conventions in the conservation field, obligations under these latter conventions may have a significant role. These may contain obligations which clarify or supplement vague or general provisions in the Ramsar Convention. Without going into detail, interactions of this kind may arise, for example, with the obligations under the Bonn Convention in particular (see Chapter 3.1.) and with the Convention on the Conservation of European Wildlife and Natural Habitats, as well as regional agreements in other parts of the world (24).

This phenomenon also exists in respect of global or regional conventions on the prevention or limitation of pollution. Conventions relating to marine pollution are of particular significance in this context.

3.4. The European Community (EC)

For Member states of the EC, obligations under the Ramsar Convention are supplemented by various Community law obligations, ranging from the Directive on Environmental Impact Assessment (25) (itself a refinement of the obligations under Articles 3(1) and 3(2) of the Ramsar Convention as to planning and wise use, etc.), to the Bird Protection Directive (26). Article 4 of the latter contains stringent obligations which, in principle, states may not disregard, to protect bird habitats, particularly those of species listed in Annex 1 to the Directive (many of which are migratory species). Furthermore, a special Council Resolution contains an obligation to designate special EC Bird Protection Areas (27). The provisions of the Directive on the Protection of Wild Birds derive a particular character from the obligation of the "independent" EC Commission to monitor compliance with and fulfilment of Community obligations, and from the potential to bring proceedings in respect of infringements of Community law before the Court of Justice in Luxembourg (28).

3.5. Conclusion

It would seem from the foregoing (especially Chapter 3.1., but perhaps also 3.2. and 3.3.) that it can be confidently asserted that there now exists a principle of customary international law that states are under an obligation to protect and conserve ecosystems within their jurisdiction, which on the international plane, are rare or endangered or have special importance as habitats for migratory species of wild animals (28a). It follows that it could therefore be argued that even independently of the provisions of the Ramsar Convention, Denmark is already under an international obligation to ensure the wise management not only of the sites designated under the Convention but also of other similar sites. It is by no means clear that the terms of the Ramsar Convention are more highly-developed than this suggested principle of customary law. Admittedly, the principle is subject to the qualification that encroachment upon such sites may be entertained on the grounds of urgent national interests, but the same is true of the obligation contained in Article 4(2) of the Ramsar Convention. Indeed, after 15–20 years of life, the Ramsar Convention now appears a little dated, even old-fashioned, in the light of present international environmental law. However, there are clear indications that, via the conferences of the contracting parties from being a passive instrument for protection the Ramsar Convention is developing into an active tool for the promotion of a sustainable development in the developing countries in particular (7).

It is in the light of these considerations, among others, that the implementation of the Ramsar Convention should be examined — see Chapters 5 and 6.

. The Ratification of the Ramsar Convention

.1. The Background to Denmark's Accession

enmark participated in the diplomatic conference held early in 1971 at which the onvention was adopted. The Convention requires only 7 ratifications to enable it enter into force. Although three times that number of states (including enmark) signed the Convention, it did not come into force until nearly five years ter on December 21, 1975. Two more years passed before the Convention ame into force in respect of Denmark, as the 19th Contracting Party — see hapter 1.

leanwhile, Denmark took part in the 1972 UN Conference on the Environment Chapter 3.1.), and co-signed a recommendation to states to accede to the Con- ention (29). Furthermore, the European Community's first plan of action on the nvironment called for special efforts to protect wild fowl and their habitats (30), nd, on December 20, 1974, the European Commission recommended that lember states, none of which had at that time ratified the Convention, accede to e Convention. In 1976 the Council of Europe launched a Europe-wide campaign r the protection of wetlands. Finally, also in 1976, a question was raised in the *olketing* (the Danish Parliament) enquiring of the Minister for the Environment as what progress had been made on Danish ratification of the Convention (31).

.2. Preparations for the Designation of Sites

itially, preparations for Danish ratification focussed on the obligation in the onvention for ratification to be accompanied by the designation of one or more tes for inclusion on the List of Wetlands of International Importance (Chapter 3.).

1974, a working group under the Danish Ministry of the Environment presented proposal for 22 Danish sites. An extensive round of hearings had to be con- ucted, so despite good intentions and the support of the relevant agencies and her authorities, it was the beginning of 1977 before a final list was ready, for esentation to the Minister for the Environment and the Danish Government with view to formal ratification of the Convention.

addition to the Ministry of the Environment, those Ministries most closely con- erned in this final phase were the Ministry of Agriculture (which then had jurisdic- on over the conservation and management of wildfowl and wildlife reserves, and hich played a central role in the designation of Denmark's sites), the Ministry of ansport (through which sovereign rights over its territorial waters are exercised) nd to some extent the Ministry of Foreign Affairs. When the precise delineation of e individual sites was carried out, factors such as planned urban, industrial and arbour development were of course taken into account — see also the Govern-

ment Memorandum and Circular from 1980 described in Chapters 4.3. and 4.7.
In addition, a number of compromises were necessarily struck in the process.

The political and administrative climate for nature conservation was then differe
to that obtaining today. Indeed, it can candidly be admitted that that there wa
considerable distrust of the Convention and the restrictions which it might impos
On this basis the result which was received was very satisfactory, both from
national and an international perspective.

The final proposal for designations involved 26 sites of a total area of almo
6,000 km^2 (56). This figure was achieved even though the most important we
land, the Wadden Sea, had to be omitted in the first instance (although at the tim
of ratification a special declaration was given in this respect) and despite certai
other preconditions attached to the ratification (32) — see below.

4.3. The Ratification Decision

The ratification resolution was passed by the Danish Government on March 2
1977. After the Ministry of the Environment had requested the Ministry of Forei
Affairs to take the necessary action, the Queen of Denmark ratified the Conve
tion by Royal Resolution on July 16, 1977 (33).

The Government's ratification resolution was based upon a memorandum draw
up by the Ministry of the Environment (34) expressing the opinion, among oth
things, that the mere inclusion of a wetland on the List does not imply an actu
obligation to place that wetland under a special conservation regime but simply
duty to manage the site (and other wetlands) in order to maintain their ecologic
character. It was for this reason that only a very modest proportion of the 26 sit
notified by Denmark were at that time subject to specific conservation regimes.

From the moment of ratification, Denmark has adhered to the view that the List
Wetlands of International Importance should not be restricted to sites which a
subject to specific conservation regimes, and it has framed its pattern of design
tion accordingly. The foundation for this interpretation is that, while Article 3(1)
the Convention onlyimposes an obligation to *promote the conservation* (author
emphasis) of the wetlands included in the List, under Article 4(1) there is a
obligation to promote the conservation of wetlands by establishing nature rese
ves, whether or not these wetlands appear in the List. This clearly implies that t
duty to designate wetlands for inclusion on the List is not confined to those whi
already enjoy the status of nature reserves. This more restrictive view has n
been put forward (see Chapter 4.6.1.), but designation policies have varied.

Designations by Canada and Sweden have been based on interpretations simil
to those of Denmark. On the other hand, countries such as the UK, t
Netherlands and Norway have chosen to designate only sites which alrea
enjoyed the status of nature reserves or something similar.

The memorandum to the cabinet session concludes "that Danish ratification will not have financial or administrative consequences requiring budget allocations, for which reason the question of ratification will not be submitted to the *Folketing* (the Danish Parliament)". This conclusion reflects the terms of Article 19(1) of the Danish Constitution which provides that the consent of the *Folketing* is not required if no action on the part of the *Folketing* is required to give effect to the obligations undertaken. This conclusion flows from the interpretation referred to above, to the effect that the requirements of the Convention could be met within the terms of the existing legislation.

4.4. The Involvement of the *Folketing* Committee

The second provision to Article 19(1) of the Constitution, to the effect that the consent of the *Folketing* is required if the obligation in question "is otherwise of great significance" is not referred to in the memorandum. On the other hand, in view of the fact that the *Folketing* Standing Committee on Physical Planning had previously been informed of the preparations for ratification in another context, an undertaking was given that the Minister for the Environment would brief that Committee on the ratification resolution and the designated sites. This took the form of a letter dated May 16, 1977 addressed to the *Folketing* Environment Committee (35).

As early as the following day, May 17, 1977, this briefing provoked a question from the Committee concerning the implications of a site being designated for inclusion on the List, *as well as* "the ratification procedure employed". These questions were answered on June 21, 1977, more or less in conformity with the memorandum to the cabinet session (36).

On July 14, 1977 the Committee raised six more questions of a very precise nature. These included whether the designation implied restrictions on agriculture, the erection of windmills for irrigation, hunting and fishing, outdoor recreation, holiday accommodation, harbours, etc. On August 31, 1977 the Minister for the Environment replied that the Convention does not imply any obligation to subject the sites to "additional legal protection" nor involve "direct legal obligations towards the citizens" or "the imposition of further limitations on the right of free disposal, nor limitations on the right of free disposal in respect of the matters referred to in the question". Furthermore, reference is made in the conclusion to the fact that the conservation and hunting authorities will endeavour to exploit the opportunities, provided by legislation, to protect ecological interests. For a more detailed analysis of this very (and undoubtedly too) prudent answer, see note 37.

4.5. The Legal Basis for the Ratification

4.5.1. Article 19 of the Danish Constitution

As referred to in Chapter 4.3 above, ratification took place under Article 19(1) of

the Danish Constitution, which confers on the Government the right to ente
international obligations which are not "of great significance" and for the fulfilmen
of which the concurrence of the *Folketing* (the Danish Parliament) is not requirec
(37a).

4.5.2. Section 60(b) of the Danish Conservation of Nature Act
Similarly, Section 60(b)(1) of the Danish Conservation of Nature Act permits the
Government to enter into agreements with foreign states for the purpose of carrying out joint activities to meet the objectives set out in Section 1(2)(1) and (2) o
the Act. These objectives include the conservation of plants, animals and areas o
significant scientific, educational or historical interest. Section 2 of the Act enables
the Minister for the Environment to make regulations for the performance o
international agreements of this kind.

This provision was added to the Conservation of Nature Act as late as in 197
(38) when preparation of the Danish ratification of the Ramsar Convention was
actually well under way (Chapter 4.1.) but the Convention is not mentioned in the
explanatory notes (38a) to the Bill introducing the amendment, despite the fac
that the scope of the statutory provision is there described as "agreements whic
cover a wide area within the objectives of the Conservation of Nature Act" (39)
Neither was Section 60(b) mentioned in connection with the ratification — se
Chapter 4.3. (40).

On closer examination, however, this is not as strange as it may appear. Sectio
60(b)(1) of the Conservation of Nature Act (compared with the express referenc
in the explanatory notes to the so-called Washington Convention on Internationa
Trade in Endangered Species of Wild Fauna and Flora (41)) is directed particu
larly at being able to conclude and implement treaties, for the implementation o
which the concurrence of the *Folketing* (the Danish Parliament) would otherwis
be required. It applies equally to treaties which are simply incorporated int
domestic law and to those which are recast in the form of an executive order (43)
As mentioned in Chapter 4.3., however, the Ramsar Convention was ratified o
the basis of so-called "establishment of the existence of harmony", i.e. the findin
that the existing law, including the discretionary powers which that law conferre
on the administration, made it possible to meet the obligations imposed by th
Convention, particularly in respect to the designated 26 Ramsar sites.

The conclusion was therefore drawn that there was no need to invoke Sectio
60(b). Nonetheless, the fact that the provision was not referred to "*ex tuto*", nor i
connection with its consideration by the *Folketing* Committee, (see Chapter 4.4.
may have been due to considerations of political tactics, particularly as the Gov
ernment further consolidated its position by involving the *Folketing* (through th
Folketing Committee) in the ratification resolution, albeit only by means of
briefing.

4.5.3. Use of the Authorization Provision under Section 60(b)(2)

The fact that there is no reference to Section 60(b) in the memorandum on the ratification resolution is hardly likely to have the effect of excluding the application of subsection (2) (authorizing the Minister for the Environment to make regulations to give effect to international agreements concluded by the Government pursuant to subsection (1)), should it be necessary to do so. The necessity may arise where the existing law is, or proves to be, inadequate to ensure compliance with the obligations in question, a matter of some significance in the Ramsar context since the Convention is continually developing, for example through "authoritative" interpretations of its provisions by the Conferences of the Parties (see Chapter 3.2). Two points should be noted in this regard. First, according to established practice, there is some discretion as to the choice of the precise manner in which the obligation is discharged (44). Secondly, it would be unreasonable to suppose, merely because the Government entered into a treaty on the assumption that the obligations imposed by it could be discharged without amendment to the existing law, that that in itself would bar the modification of that law in pursuance of an unequivocal general authority to do so in a case where such modification was necessary. In a proper case, this right to amend the existing law would amount to a duty on the part of the Minister for the Environment to take such action, and any conflict of jurisdiction in this connection would have to be settled at inter-Ministerial level (45).

4.5.4. Limits of Application — The Rule of Interpretation and the Rule of Presumption

These principles, however, are subject to certain limits. For example, it would not be possible to impose new restrictions on the right of private individuals freely to dispose of their property, a point in respect of which assurances were given to the *Folketing* Committee at the time of the ratification — see Chapter 4.4.

Furthermore, to a certain extent the application of the authorization provision in Section 60(b)(2) assumes that the underlying conflict cannot be resolved by the application of either the *canons of interpretation* or by *presumption.* The former requires that, where a rule of law is capable of two interpretations, that which is the more compatible with the terms of the treaty is to be preferred. The latter presumes that where a law is enacted subsequent to the adoption of the treaty and a conflict appears to exist between the law and the treaty, the law should be applied in such a manner as to comply (or to continue to comply) with the treaty obligations (46).

Presumably, these principles would also apply to the very question of the extent of the competence conferred by Section 60(b)(2) itself. It is therefore quite clear that the Minister *can* use his powers to re-establish harmony between the domestic law and the treaty in cases where there are no grounds for assuming that a breach of treaty was contemplated on a subsequent change in the law.

Where such a discrepancy between the domestic law and the treaty has come about as a result of administrative action, there is presumably nothing in principle to prevent the use of Section 60(b)(2) subsequently to re-establish conformity between them, although in such a case also, any problems of jurisdiction must be settled at inter-Ministerial level.

If, however the conflict derives from the enactment of a law which relates directly to a matter which the Convention is intended to regulate, then in general terms there can be no room for the operation of Section 60(b)(2). This does not necessarily mean that Section 60(b)(2), cannot be appealed to as statutory authority for the promulgation of regulations to abrogate the offending law, but only that the subsection in itself cannot be pleaded as having, without more, the effect of reversing the express provisions of that law.

In all other cases where conflicts with subsequent changes in the law deliberately conflict with the Convention, or are acknowledged to have that result, the application of Section 60(b)(2) can scarcely be excluded in advance, except in accordance with the principles mentioned above (47).

4.5.5. The Principle of Legality

The use of the powers contained in Section 60(b)(2) would, furthermore, be of particular importance if the subsequent statutory provision conflicts with the obligations under the Convention, because the Conservation of Nature Act itself does not provide for the requirements set out in the Convention to be taken into account in arriving at actual decisions in individual cases; nor does it insist that the competent authority, when operating the later provision, should have regard to the Convention requirements. In such cases, the principle of legality (48), according to which any executive act must have a basis in domestic law, in effect means that the authorities cannot fulfil their "duty" to "adhere to the requirements in a convention when interpreting the relevant legislation or exercising discretionary powers authorized by the Act" (49). The Convention does not in fact in itself provide the authority for the issuing of administrative decisions (50).

4.5.6. Practice

The powers conferred by Section 60(b)(2) have hitherto not been invoked in connection with the Ramsar Convention (not even in the circumstances described above) perhaps because there has been no obvious need. The limitations on the right of free availability of property which have been introduced subsequent to coming into force of the Convention, (and which, inter alia, reinforce the Convention's aims), have all taken the form of new legislation. This matter will be discussed in more detail in Chapter 5.3.

4.6. The Premises for the Ratification

4.6.1. The Promotion of Conservation

As mentioned above, ratification took place in the expectation that it would be possible to meet the obligations contained in the Convention by means of the existing legislation.

this connection the assumption was that the Convention did not merely imply
obligation to designate areas enjoying protected status or to protect desig-
ated sites, but to promote wetlands conservation generally — see Article 4(1)
d Chapter 4.3. While the Government could, of course, assume such an obliga-
n in principle in respect of territorial waters and government property, (see
ction 60, Conservation of Nature Act), on the other hand, the designated sites in
rritorial waters were very large. Furthermore, and more importantly, the conser-
tion (including the decision in this respect, cf. Chapter III of the Conservation of
ature Act), of areas in private ownership was the responsibility of the Nature
onservation Boards and the Chief Conservation Board, organizations over
hich the Government does not enjoy such relatively categorical powers of dis-
etion. In addition, problems would have arisen concerning the budgetary alloca-
ns for compensation payable for conservation-related purposes and the
volvement of the *Folketing* under section 33, Conservation of Nature Act (50a).

6.2. Planning Act Reform and the Environmental Protection Act

e Government was, however, well placed to tackle the obligation to protect
etlands, manage them according to sound ecological principles and avoid
lverse changes to or destruction of wetlands — see Chapter 2.3. An extensive
form of the planning law had been carried out, furnishing the central authorities
th a good grasp of the underlying principles of planning and the content of the
ucture plans, as well as of decisions concerning virtually all categories of
anges of use of land — particularly as "instructing" and appeal authorities. A
nilar degree of control already existed over territorial waters, by virtue of the
vereignty of the State over such waters. Even in the field of pollution control, the
ntral government at that time probably assumed that with the help of the then
atively recent Environmental Protection Act (which had been described as "the
st in the world"), improvements would be discernible in the foreseeable future,
rticularly in respect of the pollution of fresh water and in coastal zones. At least,
would have been expected that any deterioration in the situation could be
oided.

6.3. Agricultural Use

ere was one area, however, where it was plain that optimal fulfilment of the
nvention's obligations could not be achieved without changes in the law,
mely in relation to changes in general agricultural use. Presumably, the same
s also true in principle for fisheries. A contrasting view held that the Convention
not impose any duty to modify the legal regime governing the existing, lawful
e of land (see e.g. Chapter 4.4.). Moreover, there was presumably also an
pectation that it would probably be possible to prevent the worst "accidents", by
r example) conservation orders under Chapter III or Section 60 of the Conser-
tion of Nature Act.

6.4. Limitation of the Discretionary Powers of the Administrative Authorities

mentioned in Chapter 4.3., at the date of ratification, the central obligation of
Convention was thought to be to ensure that the 26 designated Ramsar sites
particular were managed according to sound ecological principles. Specifically,

designation indicated cases where the ecological value of a site was of a nation (and thus international) character, so that conflicts of interest could arise only i the form of conflicts *with other national considerations* (34). This was seen a significantly limiting the exercise of discretionary powers in general, and as affec ing actual executive decisions concerning the disposition of land, since in this wa the exercise of discretionary powers was in principle restricted to conflicts be ween different classes of national interest. On the other hand, the concept designation was already familiar, having been employed for several years i identifying (for the benefit of authorities engaged in deciding individual land-us questions) areas whose conservation and protection was of national importanc (51).

The new feature was that designation which took place formed an external co straint which was naturally bound to influence the central authorities in the exe cise of their powers over subordinate authorities. Designation ensured initial co pliance with the Convention, but it was naturally important to communicate th message that these were Ramsar sites and to explain their significance, so as ensure that the necessary steps were taken to respect the Convention's requir ments on a day-to-day basis in the actual practice of the authorities applying th law (52). This task was not a simple one, for the Convention covers a large are not only in a geographical sense! Many authorities and many statutes, other leg instruments, circulars, etc., have a role to play in planning and land use, and ma of the authorities each possesses its own particular area of jurisdiction, whic never makes a task of this character any easier.

4.7. The Implementation of the Convention in Denmark

4.7.1. The Proclamation of the Convention

The Convention entered into force for Denmark on December 19, 1977. This ga rise to an obligation to interpret the existing law and exercise administrati powers (including the making of decisions) in accordance with the terms of t Convention, particularly in relation to the designated 26 Ramsar sites (53). In t context of this paper, it can only be of academic interest (54) whether this effe occurred automatically on December 19, 1977 or was dependent on the Minis of Foreign Affairs' proclamation of the Convention in the Official Gazette C April 4, 1978, with retrospective effect from December 19, 1977. However th may be, the proclamation drew the attention of the relevant authorities to t Convention and the obligations implied by it, particularly since the proclamati contained a description illustrated by maps of the 26 Danish wetlands which h been notified as being of international importance. At the same time, all t competent authorities and courts were put on notice of their duty to apply t Convention and to respect its obligations (55).

4.7.2. Other Matters Relating to the Convention

As mentioned in Chapter 4.2. the Ministry of Agriculture and the Ministry of Tra port (at that time the Ministry of Public Works) were directly involved in t preparatory work for the ratification of the Convention, and the views of, amongst others, the Association of County Councils in Denmark had been heard in this connection. Therefore the Danish authorities most closely concerned had first-hand knowledge of the Convention and its effects.

Furthermore, the Convention had been described to all municipal authorities in Circular No. 124 of June 16, 1977 on the planning of holiday cottage districts, in which it was stated, even before the Convention had come into force for Denmark, that maps showing the 26 nominated Ramsar sites "would be included in the proposal" for the designation of those coastal areas in which holiday cottage construction would not be countenanced. This designation was subsequently extended by Circular No. 167 of August 28, 1981 (which replaced the earlier circular) so as to apply to the planning of holiday and leisure accommodation generally, e.g. to hotels and camping sites also. In the context of the regional planning legislation, this amounted to a national planning directive; in other words, the provisions of the circular should be taken as binding on the competent planning authority (56). This was a very "determined" fulfilment of the Convention requirements, even before it had come into force, and thus also served as an efficient information channel.

Furthermore, on October 6, 1977, all counties and the Greater Copenhagen Council individually received a copy of the text of the Convention trnslated into Danish, the maps and the supporting documents for the government's ratification resolution (in the form of the Minister for the Environment's report to the Folketing Committee of May 16, 1977, see Chapter 4.4. and note 35). By a letter dated November 24, 1977, some of this material was also sent to the then-existing Conservation Planning Committees which included local authority representatives.

4.7.3. The 1980 Circular on the Convention

As a result of all this activity, having regard to both the coming into force of the obligations and the information and instructions about them, it was probably somewhat "*post festum*" for the National Agency for Protection of Nature, Monuments and Sites to issue on September 1, 1980 a circular (Circular No. 138) about the administration of the Convention, although in addition to regional and municipal authorities the circular was also addressed to the Nature Conservation Boards and the Chief Conservation Board. The circular's significance, however, is that it reports on the new legislation which had been passed in the intervening period. In addition, the circular presumably also arose from a desire to improve the mechanism for complying with the obligation in the Convention to inform the Convention Bureau of any actual or anticipated deterioration in the notified Ramsar sites (see Chapter 2.3.), by imposing a duty on the authorities concerned to notify the National Agency for Protection of Nature, Monuments and Sites of any such matters (item 5 of the circular). The necessity to do so was heightened by the imminence of the First Conference of the Parties (Cagliari, November 1980). Finally, the circular also contained some administrative definitions (57) which had become necessary as a consequence of administration of the Convention over the intervening 2–3 years. The circular thus may actually have had a law-making ("quasi constitutive") effect (58).

4.7.4. Moreover

The Ramsar Convention has naturally been mentioned or given general prominence in a number of other respects, including articles, books, reports, guidelines, etc. (59). A small selection of these will be mentioned in the following chapters (1).

The island of Nekselø and the surrounding sea area in the Sejerø Bay was notified as a Danish Ramsar site (no. 18) in 1978.

5. Ongoing Implementation of the Convention

5.1. Introduction

5.1.1. Outline

The Chapter describes the manner in which the Convention has been put into practice in the broadest sense, that is to say how the Convention has been implemented on a continuing basis. The intention, however, is not to give an exhaustive account, as the matter is too complex to make that possible. Treatment of the exercise of discretionary powers in individual cases will therefore deal by and large only with the practice of the central authorities. In the planning field, only final planning decisions will be described, and not the reports and similar background documents on which these decisions were based. References to administrative precepts and guidelines will also be incomplete. Finally, there is no clear delineation between on the one hand the steps taken to implement the minimum requirements of the Convention and on the other those which have been taken with a view to improving so far as possible the achievement of all the Convention's objectives and requirements, perhaps in the light of developments in customary international law subsequent to the drafting and entry into force of the Convention, see Chapter 3.1. New legislation provides only a limited guide to this boundary, although it can quite logically be argued, on the basis of the ratification decision, its background and assumptions, that the obligations extended no further than was permitted by the then existing law.

In view of what follows hereafter, it is also important to recall that the Convention itself specifies virtually no particular measures for its implementation. The method chosen is thus left to the discretion of each Contracting State — see Chapter 2.4..

5.1.2. Subsequent Legislation and the Rule of Interpretation

The issue has also been blurred by the fact that the purposes of the Convention have also been furthered by subsequent legislation, even though in certain cases such legislation was aimed at achieving a particular (but slightly different) objective. A typical example is the introduction in 1978 of general provisions as to conservation of bogs of at least 0.5 hectare in extent — Section 43, Conservation of Nature Act. This provision was not enacted in fulfilment of the Convention requirements, but it did enhance the general level of conservation for wetlands. Therefore, it was not to be wondered at that the 1980 circular mentioned in Chapter 4.7.3. proposed a more restrictive application of the new rule to bogs lying within the 26 designated Ramsar sites (60). It is more than doubtful whether this can be justified by reference to the rule of interpretation (see Chapter 4.5.4.), which favours that interpretation of new regulations which renders them most closely compatible with the obligations of the Convention (concerning ecological management, particularly in designated areas) and with the principles of international law generally. The instituting of a general regime of conservation can hardly be said to be required as a consequence of these rules, far less to be in conflict with them. It may also be questioned how far a more restrictive practice is a logical consequence of a single matter being designated as internationally important in a legal instrument.

5.1.3. The Significance of Development
At any rate, this does show the difficulties from allocating different measures to one particular category, as well as the importance which development can achieve on both national and international levels, even when the starting point is a static one: one particular convention, of which the obligations can be fulfilled on the basis of one particular current legal order. On this basis, a variety of legislative provisions and administrative measures are here included under the heading of the ongoing implementation of the Convention, even though they would not normally be regarded in that context.

5.2. Supplementary Rule of Interpretation?

Actually, where legislation produces a regime which is stricter than that whicl formerly existed the above example also illustates that in cases where legislatior tightens the former legal order, or makes that regime more rigorous in pursuanc of the overall objective of a convention acceded to on the basis of the establish ment method ("harmony of norms"), it is appropriate also to consider the effects of that legislation within the Convention's sphere of operation (see also Chapte 5.3.3.). If fulfilment of the obligations of the Convention originally required a more stringent practice within the Convention's sphere of operation than that generally applicable, is it then the legislator's intention that the only change should be a substitution of the Convention standard as the basis for a uniform practice withi the scope of the entire Act, or should practice be tightened further, so that there continues to be a "difference in level" between those matters covered by the Convention and other matters? This problem is particularly relevant within the area of environmental legislation, where the trend is towards continued tightenin of legislation. The rule of interpretation does not appear to solve the problem However, it may be that closer examination of the entire set of environmenta rules in the context of all the applicable international obligations will show that i practice a supplementary rule of interpretation has developed within this area, i which importance should be attached, not only to the direct obligations in the Convention, but also to the Convention's broader purpose. This could be (further based on international "soft law" in the area (Chapter 3.3.) and the developmen of customary international law (Chapter 3.1.), etc., as well as on the underlyin national interests in achieving more effective conservation of the global environ ment, and thus also an eventual improvement in conservation of the nationa environment, through a tightening of international law obligations relating to the environment. However, a survey of this nature lies beyond the bounds of this report. My own cautious evaluation is that the result of the survey would be positive to a certain extent.

It is also possible that the investigation might prove that development is influ enced by efforts to formulate conventions in terms soft enough to permit of acces sion by as many states as possible. Language such as "to promote the conserva tion" (of sites) (see Chapter 2.4.) creates an obligation which *does* have a certai energy or dynamics. While it is not clear and unambiguous, it is so cast that it ca itself influence development in national legislation. Conversely an understanding of the inherent scope of the obligation in the Convention can be influenced by national law.

5.3. Legislation

5.3.1. Section 43 etc. of the Conservation of Nature Act

The most important "implementation measures" in this respect are amendments to i.a. Section 43 etc., Conservation of Nature Act. Amendments introduced in 1978 (Act No. 219 of May 24, 1978) improved the protection of watercourses and lakes, as well as general conservation of bogs of at least 0.5 hectares, and further amendments in 1983 (Act No. 208 of May 25, 1983), introduced general protection of salt meadows and salt marshes of over 3 hectares and of heathlands of over 5 hectares (61). In this connection, it should be noted that Act No. 250 of May 23, 1984 reduced the maximum size of lakes falling under Section 43 of the Conservation of Nature Act from 1,000 m^2 to 500 m^2. The provisions of Section 43, etc., imply that no changes may be made to these ecosystem types without the authorization of the County Council/Greater Copenhagen Council, with (now) the National Forest and Nature Agency as the appeal authority.

The circulars made subsequent to these amending Acts do not mention the Ramsar Convention in particular, but as mentioned in Chapter 4.7.3., general Circular No. 138 of September 1, 1980 states that the National Agency for Protection of Nature, Monuments and Sites (now the National Forest and Nature Agency) will allow consideration of the Ramsar sites "to be given considerable weighting" in treatment of appeals under the provisions of the 1978 Act on the protection of wetlands. This principle has subsequently been repeated — in a slightly different and perhaps more general form — in a memorandum of July 1, 1987 from the National Forest and Nature Agency, approved by the Minister for the Environment, concerning the establishment of deer farms, published in the current information letter to i.a. the authorities involved, entitled "Section 43 NEWS" (entry no. 41). From this it appears that in rulings, which are in principle discretionary, on whether interventions in i.a. wetlands should be authorized, the exercise of "discretion is restricted ... by international conventions such as the Ramsar Convention" and that it would be "incompatible with the primary interest of conserving these nature areas if authorization were granted for the establishment of deer farms". There is no reason to believe that other essential interventions will be treated differently from deer farms. The extent of this restriction should be compared with the fact that it appears from the Minister for the Environment's report of January 11, 1985 (question no. 129 — general section Annex 146) to the *Folketing* Environment and Planning Committee that, of over 1,000 km^2 of land comprised in the (then) 26 Danish Ramsar sites, probably 400–800 km^2 fall under the Section 43 system of the Conservation of Nature Act (62).

In the context of the Conservation of Nature Act, the recent Act on the conservation of the outer marshlands in the Tønder Marsh (Act No. 111 of March 12, 1988) should also be mentioned. To a great extent, the Act regulates the agricultural use and farming of the area, with a view to (see Section 2) preserving the outer marshlands, etc., of the Tønder Marsh "as a consolidated nature area of national and international importance". In Section 2 of the explanatory notes to the general introduction (Bill No. L 77, p. 15) reference is i.a. made to the 1987 designation of the Wadden Sea, with its adjacent land areas, as a Ramsar site (32). The conservation objective of this Act, as well as its regulatory provisions (to a certain extent), do not differ substantially from the concrete conservation measures described in Chapter 6.2.

5.3.2. The Water Supply Act and the Marine Environment Act

Under Section 2 of the Water Supply Act (now Consolidated Act No. 337 of July 4, 1985) i.a. nature preservation must be taken into account in the administration of the Act. From the explanatory notes to the Act of 1978, from which Section 2 derives its present wording, it appeared, for example, that water abstraction can be completely prohibited in certain areas, due to the ecological consequences. In this connection, the 26 designated Ramsar sites are described as "wetlands requiring special protection" (63).

Furthermore, the Ramsar sites are mentioned in the explanatory notes to Chapter 2(3), and Section 34 of the 1980 Act on the Protection of the Marine Environment (Act No. 130 of April 9, 1980, as amended by Act No. 181 of May 8, 1985) as areas for possible special regulations on the discharge of sewage, etc. by pleasure vessels and on the prevention of marine pollution by substances not falling under the Act's general prohibition (64). However, these powers have so far not been exercised. On the other hand, in some cases, under Section 60 of the Conservation of Nature Act on conservation regulations in territorial seawaters (of e.g. the Ramsar sites), stricter rules have been laid down in these respects than those immediately in force according to the Marine Environment Act (65).

5.3.3. Other Environmental Legislation

In both of the above cases, it can perhaps be said that the application of ordinary principles of interpretation might produce the result shown in the explanatory notes. However, when legislation is implemented after a relevant convention has come into force in Denmark, it is natural to refer to the convention in relevant contexts, so that the future legal regime can be defined. If this does not take place, uncertainty might arise, cf. Chapter 5.2. As far as can be seen, however the Ramsar Convention is not directly referred to in connection with other environmental legislation. There are, however, more general references to "international obligations" or similar (66) in the explanatory notes to certain other Acts. This is so, for example, for the Act on the Environment and Gene Technology (Act No 288 of June 4, 1986) where the reference is made as a justification for controlling to some extent activities within the national fishing zone, under Section 4 (67). To the extent to which environment and land-related legislation is involved, the Ramsar Convention will presumably be included under these or similar references.

5.3.4. The Hunting and Game Management Act
There are no similar references in the explanatory notes to the 1982 extension of the Hunting and Game Management Act (now Consolidated Act No. 297 of June 6, 1984) to apply it also to the fishing zone in respect of i.a. conservation provisions (Section 56a). This extension is presumably also based on improved fulfilment of the objective of the Ramsar Convention, since not all of the designated areas lying outside the base line lie entirely within territorial waters.

5.3.5. Agriculture Legislation
Agriculture-related legislation raises special problems, since in some respects it involves more or less automatic statesubsidies for e.g. land drainage, irrigation, shelter hedges, etc., which directly conflict, or might conflict, with the objective of the Convention.

This matter was not directly considered on accession to the Convention (cf. Chapters 4.4. and 4.6.3.), so that it must be assumed that it was not thought of as being in conflict with the Convention obligations. Furthermore, the unfortunate consequences, from the point of view of the Convention's objective, have to some extent been remedied by the tightening of the Conservation of Nature Act described in Chapter 5.3.1. Taken as a whole, *this* is perhaps the explanation for no effective measures apparently having been taken to remedy a situation where the Government is, on the one hand, committed to wise ecological management of at least the notified Ramsar sites and, on the other hand, provides agriculture with subsidies for activities tending to the contrary result.

The matter was improved by the 1985 amendments (by Act No. 56 of February 20, 1985) of the Act on Subsidies for Drainage and Irrigation, which removed state subsidies for the draining of certain meadows (Section 5(b)(3) concerning "permanent grasslands which have not been regularly redesignated").

Furthermore, according to the explanatory notes to Section 18 in the Hedgerows Bill (No. L 136) put forward by the Minister for Agriculture on January 13, 1988, subsidies for hedgerows "will be refused where the establishment of hedgerows would be in conflict with ... obligations resting on Danish authorities within the framework of international agreements such as the Ramsar Convention ..." (67a).

5.3.6. Agricultural Development within the EC
Agricultural development within the EC (in particular set-aside) naturally exerts a positive impact in relation to the objective of the Ramsar Convention. At the time of writing a number of initiatives, including Bills to strengthen conservation, are under way. These also include more sophisticated methods of fulfilling objectives of the Ramsar Convention, such as improved supervision and monitoring — see Articles 3(2); 4(1) and 4(3) — as well as the possibility of restoring damaged areas (cf. Articles 4(2) and 4(4)) (68).

5.4. Administrative Regulations

5.4.1. Review

This section will, for the most part, describe some of the administrative regulations affecting private individuals, since regulations concerning subordinate authorities which are not described above or directly concern the regulations here under review largely deal with physical planning only, a matter which will be considered in Chapter 5.5. To a certain extent, however, various other regulations will be considered in Chapter 5.6. concerning administrative practice. As previously indicated, the division of Chapter 5 is not particularly rigid and is generally founded more on practical than systematic considerations.

5.4.2. Territorial Waters

According to Executive Order No. 489 of September 28, 1981 issued by the former Ministry of Public Works, bathing jetties and landing stages as well as "other fixed installations" (see the circular from the same Ministry No. 13 of January 31, 1984) may not be located in territorial waters without a special permit. This provision is a consequence of the sovereignty of the state over territorial waters. Its implementation includes considerations related to nature conservation, particularly in the Ramsar sites, on the grounds that protection of nature must here be considered a significant national interest (Chapter 4.6.4.). Therefore, this circular stipulates that, in discharging their functions under the executive order in permitting small bathing jetties and landing stages, the county councils and the Greater Copenhagen Council should refrain from granting such permits in Ramsar sites (Section VIII of the Circular). The same principle presumably applies with respect to breakwaters and coastal protection systems (69).

In this connection, the Ministry of the Environment has decided on a number of occasions, in the context of the approval of regional plans, that dinghy berths and launching jetties for small boats, like pleasure boat harbours generally (cf. Chapter 5.6.2.), are a regional planning matter when they are located in Ramsar sites (70).

The establishment of salt-water fish farms requires the authorization of the Ministry of Fisheries, see the Salt Water Fishing Act (Act No. 306 of June 4, 1986), Section 14, according to which relevant authorities (including the Ministry of the Environment), etc., must be consulted on applications for salt-water fish farms. Under the Marine Environment Act the operation of such activities, however, requires the authorization of the National Agency of Environmental Protection, as a precondition to the grant of a permit by the Ministry of Fisheries. According to Section 4.2. of the Agency's "provisional informative briefing on the approval of the location and operation of salt-water fish farms" (August 21, 1980), in order to avoid deterioration of water quality, "permits will not normally be given for the establishment of salt-water fish farms in i.a. Ramsar sites" (71).

.4.3. Other Regulations

'y virtue of the Ministry of the Environment's Executive Order No. 784 of lovember 21, 1986, the use of lead pellets, etc., in hunting is prohibited in lamsar sites and is also prohibited for skeet shooting (shooting ranges), if such se involves the pellets falling on certain designated wetlands, by and large efined in accordance with the Ramsar Convention definition; see Chapter 2.2. 71a).

;oncerning air traffic, the Air Traffic Directorate has stipulated (BL 7–16 of April 1, 984) that the minimum flying altitude over particularly sensitive nature areas, icluding most Ramsar sites, is 1,000 feet, whereas the altitude over other unbuilt reas is at least 500 feet.

hese provisions presently affect only the 26 original Ramsar sites and therefore o not apply to site no. 27, the Wadden Sea — and certain adjacent land areas :f. note 32 and Chapter 6.3.).

.5. Regional, Sectoral and Municipal Planning, etc. (71b)

.5.1. National Planning Directives

he Ramsar Convention, and particularly the notified Ramsar sites, have not een the subject of independent national planning directives, but they have been icluded as one of the factors to be taken into account in the planning process, ee Chapter 4.7.3. (and notes 37, 56 and 60). The Convention has also been icluded (as assumptions) in the regional planning guidelines, of which Guideline lo. 3 (from the National Agency for Physical Planning, 1978), is particularly elevant in this respect and is partly "in the nature of a circular". Although it can ardly be formally considered as a pronouncement of a national planning author-y, it is an expression of state control of the planning process itself, which is of nportance in this context (72).

.5.2. Regional Planning

he relevance of the Ramsar sites for regional planning is also touched on above, see Chapter 5.4.2. and note 71). A systematic review of all regional plans will not e attempted here, although some examples will indicate the general scope of the onvention's significance in this respect:

ı the Regional Plan 1985–97 for the County of Northern Jutland (p. 38), it is set ut as a general key objective that, within Ramsar sites (and EC Bird Protection reas) in municipal and local plans, no "action may be taken which conflicts with ıe *objectives* and provisions of the Convention, (author's emphasis), which can nly be disregarded to accommodate national interests".

he Regional Plan for 1985 of the County of Ringkøbing states in the guidelines ır meadows, heathlands, bogs, sand dunes and beaches (p. 65) that the Ramsar tes must appear in municipal plans and the relevant sector plans, with the stipulation that they may not be allocated for purposes which may make it difficult t promote their protection and wise use. There are similar guidelines for fiords (p. 70 and lakes and watercourses (p. 72). On the adoption of the regional plan, th Minister for the Environment furthermore refused to designate land for wind farm at Nissum Fiord, referring to the area's location in a Ramsar site (73).

In the Regional Plan 1985–96 for the County of Storstrøm, the Ramsar sites ar included in the special conservation areas (p. 55 of the report) governed by guideline (4.22.3.) which requires that "attempts must be made to modify installa tions or activities which have a negative impact on conservation values", and tha new installations which do not meet this requirement "must be allocated to othe areas" (p. 73).

A very detailed regional planning document has been issued by the County c Southern Jutland (February 1987) concerning the 1987 designation of the Wadde Sea with certain adjacent land areas as Ramsar site no. 27 (see notes 32, 62 an 65). The document has subsequently been incorporated in Appendix No. 3 to th Regional Plan 1985–96 (September 1988) and contains detailed guidelines fc almost all important categories of measures, as well as a number of recommenda tions to central-government authorities to contribute to the protection of the site (fc example, by refusal of subsidies — see Chapter 5.3.4.) (74).

The section of the Wadden Sea which is situated in the County of Ribe, with certai adjacent coastal areas, was incorporated in the regional plan as a Ramsar sit (under special conservation areas) as early as the 1985–96 Regional Plan (Jun 1985). As in the regional planning document from the County of Southern Jutlan (p. 80) it is stated that no dispensation can be expected from Section 43, etc Conservation of Nature Act (see Chapter 5.3.1.). In this case, however, the refe ence is not in the form of a guideline, but in the form of an undertaking, as in th regional planning document. Similarly, the Ministry of Agriculture has bee requested not to grant subsidies for projects which involve or envisage the lowerin of the water table. It is furthermore stated that permits under the Urban and Rura Zones Act cannot be expected for the establishment of large technical installations material changes in land use and the use of buildings for purposes other tha agriculture. In the explanatory notes, it is indicated that in the administration of lan use and environmental protection legislation generally, essential consideratio should be paid to the special international status implied by the designation (75)

No detailed review has been made of all regional plans and their appendices. A the regional plans are an expression of an interplay between certain planning related assumptions, various governmental reports and an approval procedure and as the Ramsar Convention obligations have clearly been taken into account i more or less all phases, it must be generally assumed that allowance has bee made for the Ramsar sites in regional plans. For this reason, either in their own rig or by virtue of inclusion in special conservation or similar areas, the sites have bee marked as areas with a special protection status, to be respected in connection wi the most significant categories of altered land use, at least.

5.5.3. The Significance and Interpretation of the Regional Plans: Veto Obligation Over Local Plans?

As the discretionary rulings of municipal authorities are governed by the content of the plans (76), the regional planning is naturally of significance for the protection of the Ramsar sites. This point is further emphasized by the fact that the regional plans are by and large binding on municipal plans and sector planning (77). On the other hand, this has in practice not proved adequate to handle all problems.

Firstly, regional plans are in fact not static. Planning is "fluid" (78) so that the plans are continually supplemented by appendices, etc. giving the main content of sector planning and/or plans for special areas, e.g. concerning wind turbines (73) or for measures which are not in accordance with the current regional plan (79). The attitude of the central authorities to these planning proposals does not differ significantly from the treatment of individual cases. The boundary between this Chapter (5.5.3.) and Chapter 5.6. concerning practice in individual cases is therefore not sharply delineated.

Secondly, sometimes problems have arisen because of lack of detail, including the exercise of discretionary powers which may be called for in the regional planning guidelines, perhaps particularly in an area such as this, thus raising the question of whether the project corresponds to the plan or not. Relations with municipal plans can also be a problem (80). Even disregarding the question of the extent to which bodies like the National Forest and Nature Agency, (possibly to a certain extent also as the appeal authority) are bound by the general obligation to accept a local plan or a project which corresponds to the regional plan (81), doubt can still arise.

However, the question is whether some of the problems or doubts which have existed have been founded on an inadequate appreciation or misunderstanding of the scope of the Ramsar Convention in this context. Neither the planning process nor the plans as such are outside the ordinary rules or principles of administrative law (82). Nor is there any reason whatsoever, in the context of international or administrative law, to consider them differently, or with greater respect, than rules of law or administrative regulations. This means that the usual principles (Chapter .5.4.) should also be applied here, i.e. both the rule of interpretation and the rule of presumption. As all the "actors" are public authorities (82a) and it is furthermore difficult to see in what manner doing so would conflict with the principle of legality (Chapter 4.5.5.), it is all the more imperative that these principles should apply.

If these principles are to be applied in the planning field, this must in general mean that, in any case where a doubt exists, the plan should be interpreted so as to harmonize as far as possible with the obligations under the Convention (the rule of interpretation). The same must also apply to administrative measures in connection with the final adoption of the plan. Even a plan which is in conflict with the obligations should presumably generally be applied in such a way that the obligations are respected (the rule of presumption). There must be an extremely stron basis (perhaps an express resolution to this effect from the *Folketing* or th Government) to justify a departure from this result.

The application of these principles will in some cases lead to a reaction against local plan, because it does not accord with the manner in which the regional pla is to be interpreted or with what it is presumed to mean — the discussion abov on the rule of presumption. In such cases, it must be assumed that the Nationa Forest and Nature Agency at least, as the authority responsible for the Ramsa Convention, is under a duty *to lay down a veto* under Section 26, Municipa Planning Act and that the Minister for the Environment is obliged to uphold th veto. The local plan would be in fact in conflict with the law, which in itself implies duty to react. However, another interpretation may be that the discretionar power, to *be able to* object under Section 26 of the Municipal Planning Act, i overridden by the Convention obligations, so that the right of protest cannot b legally ceded. The right is thus transformed into a duty.

Whether the National Forest and Nature Agency is under a duty to object to local plan "in conflict with Ramsar", but based on an adopted regional plar depends on the circumstances, but there is no doubt that there are certain limits For example, if the local plan envisages a project which is in conflict with th Convention, but founded on an approved addendum to the regional plan exclu sively relating to that project, the only remedies available will be political, i.e beyond the jurisdiction of the National Forest and Nature Agency. A hearing in th courts might be possible in theory, but will in reality depend on starting to bring legal action.

There will, presumably, be no clear delineation between cases where an objectio is based on the conflict between the local plan and the law embodying the obliga tions under the Convention, and those where an objection is made because th local plan is considered inadvisable in relation to the objectives of the Conventio (82b).

This question has not arisen in practice in its pure form. In most cases of objec tions to local plans (see Chapter 5.6.2. — 5.6.4.) the regional planning guideline have been so accommodating or inexact that there have been no major problem in taking action against a local plan which is in conflict with the Convention. Bu the action against the local plan itself, which assumes accordance with the reg ional plan, is in reality an expression of the application of the above principles b the National Forest and Nature Agency to the regional plan, i.e. "explaining" o "assuming" away any incompatibility with the Convention obligations.

Furthermore, obligations under international law are not clear and their scope i determined by authorities who are also parties to the dispute.

Nonetheless, there is no doubt that there have been cases where, in central an regional administration, any problems have been evaluated by reference to th traditional rules of the game" for public authorities in their mutual relations, and not as described above. Confidence in planning and its "political" function as a kind of beneficial act of administration may also have a role to play. Indirectly, these conclusions are, however, confirmed by the argument in the case described in note 98, for example. Why otherwise should it be pointed out that this project was not in conflict with the obligations of the Ramsar Convention?

5.5.4. Conservation Planning

Under the amendments introduced by Act No. 355 of May 13, 1987 to, inter alia, Chapter IV of the Conservation of Nature Act in respect of conservation planning, conservation plans are no longer formally binding on the regional conservation authorities, but shall merely serve as guidelines for these authorities. Yet the Minister for the Environment can continue to establish conservation planning assumptions, with which plans and individual decisions conform, just as it is still the case that neither regional nor primary municipal authorities may decide on action which is in conflict with the regional plan. In the introductory notes to the bill, EC Bird Protection Areas, and thus by implication Ramsar sites, are cited as an example of these assumptions (see notes 26 and 62).

This reflects the legal regime prior to the 1987 amendments to the Act since the Ramsar sites amongst other matters are described as binding assumptions in the guidelines and circular on conservation planning (83). Although approved conservation plans are not widely available (despite the fact that the amendment to the Act revokes the requirement for conservation plans to be approved by the Minister for the Environment, the so-called first generation plans must nonetheless be approved) and although they will no longer be binding in future, this will hardly influence the relatively strong position of the Ramsar sites.

5.5.5. Other Sector Planning

The other sector planning, which incidentally also to a wide extent falls under the amendment to the Act, is hardly of great significance. However, it does appear, from the matters discussed in Chapter 5.5.2, that to a certain extent the regional plans concerning Ramsar sites also make requirements of other sector plans. This is particularly reflected in *recipient quality plans*, since the Ramsar sites, in accordance with the general guidelines of the National Agency of Environmental Protection, are designated as areas with more stringent objectives (scientific objectives) (84). Moreover, the regional plans can indirectly have evaluated interests in advance, by highlighting the Ramsar sites' conservation status in the guidelines and its amplification in another context, e.g. that *extraction of raw materials* must take place in areas designated for that purpose, so the possiblity that future sector planning will conflict with conservation of the Ramsar sites is virtually excluded beforehand. Furthermore, a permit for raw material extraction is subject to a number of other regulations, also in individual cases. For example, according to the Conservation of Nature Act, etc., cf. the list in Sections 11–14 of Circular No. 162 of June 24, 1986, concerning the working of the Raw Materials Act (85).

5.5.6. Marine Areas
To a certain extent, i.a. in relation to recipient quality planning and *recreationa* *activities planning* (also an element of conservation planning), marine areas ar included in regional planning. But there is no actual summarized planning fo these areas, where most Ramsar sites are situated. On the other hand, under th auspices of the National Forest and Nature Agency, continual *mapping of rav material interests* takes place, supplemented with details including conservatio interests and biological interests on the seabed. Among the purposes is to enabl these interests to be taken into consideration in the designation of new areas fo mineral exploration. The mapping simultaneously contains further details of th relevant Ramsar sites and at present covers approx. 22,000 km^2 (86).

The *Folketing* Environment and Planning Committee has paid particular attentio to the question of *exploration and production oil wells* in Ramsar sites an elsewhere and the Minister for the Environment has therefore on several occa sions submitted reports on this matter. No binding statements have been made o the abandonment of these activities in principle, except that immediate respect i to be accorded to Ramsar sites *subject to conservation* (cf. Chapter 6.3.) and a statement that in the event of an unacceptable risk to the waterfowl stock n exploration drillings should be made in Ramsar sites and EC Bird Protectio Areas, and that these guidelines also apply broadly to production drillings, (87).

5.5.7. The Municipal Plans
are not of any great interest in this context, since the basis for the management o open landscape is normally the regional and sector plans, and the authorities fo them are also generally responsible for deciding individual cases regarding th open landscape. Conflicts in relation to the Ramsar sites usually arise in connec tion with these individual cases or at the local planning level (88). In this respec reference is made to the section below.

5.6. Practice in Individual Cases

5.6.1. Delineation
Considering the relatively large size of the Ramsar sites and the fact that they ar frequently located in attractive landscape areas, the individual cases comin before the central administration, including the Central Appeal Boards, are rela tively few. This is also remarkable in view of the power of voluntary "green organizations or their local sections in many cases to appeal against decision made in pursuance of the various regulations applicable. All in all, this indicates a decentralized administration which generally respects the Ramsar sites. On th other hand, the number of cases in which the central authorities become involve is nonetheless large enough for it to be possible to outline actual practice in thi respect.

The overall framework for practice in individual cases (in the form of circulars an similar documents) is described in Chapters 4.7., 5.3. and 5.4. in particular. Thi amework naturally also includes planning, see Chapter 5.5. Practice in individual ases (concerning delineation, see Chapter 5.5.3.) is to some extent also there escribed. See Chapter 5.3. in particular as to Section 43, etc., of the Conserva-on of Nature Act.

ndividual cases decided at central level have concentrated particularly on yacht arinas and hotels, as well as wind turbines (the latter both under the Urban and ural Zones Act and as local planning cases). Others have concerned a variety of ifferent installations and activities such as dumping, sightseeing flight routes and hooting ranges, etc. (89). A special group of cases concern the rules under the onservation of Nature Act, administered by the nature conservation boards and e Chief Conservation Board.

.6.2. Yacht Marinas

ocal plans concerning yacht marinas considered to be in conflict with both the amsar obligations and regional plans will normally mean that the National Forest nd Nature Agency will enter an objection or veto under Section 26, Municipal lanning Act (90). Furthermore, if the question is (or was not at that time) addres-ed in the regional plan, emphasis is attached to whether the facility is in a hitherto ntouched locality, which according to the circumstances would justify an objec-on (91); or whether it has been established in conjunction with existing facilities n which case permission might be given if the facility is on a modest scale) (92) nd/or established as an element of overall planning for pleasure boats in the elevant Ramsar sites, or on condition that such an overall plan be drawn up (93); r whether the facility constitutes general rehabilitation (94). In such cases, per-ission may be granted on the stipulation that the development will be monitored ith a view to imposing traffic restrictions either under the provisions of Section 60 f the Conservation of Nature Act relating to ministerial conservation orders (see hapter 6.3.) or under the provisions of the Hunting and Game Managment Act elating to wildlife reserves if the impacts resulting from the expansion prove to xceed the anticipated level (95). It does not appear to be very important whether e facility is established within a Ramsar site itself or merely in close proximity ereof, as it is the disturbance which is crucial. Limiting the proposed number of erths, or reservations in this respect, will therefore often be appropriate (96), hile the fact that the boats which will use the new berths already use the area ay tip the scale in favour of acceptance (97).

iewed generally, this fairly liberal practice is presumably due to the fact that everal Ramsar sites lie in obvious recreational areas, that the National Forest nd Nature Agency also handles recreational issues and that the disturbance is ot necessarily very great and can sometimes be regulated subsequently if prob-ms should arise.

.6.3. Hotels and the Like

ave a longer history of subjection to special planning requirements, cf. Chapter .7.1., and they are also included as a separate topic in the supplements to the egional plan for the planning period before 1997. However, it has not been possible to avoid problems compeletely. These have rapidly taken on a marked political character and have at times stirred up extensive public debate, in a fe instances in the period up to the supplements to the regional plans mentione above (98), which presumably by and large respect the Ramsar obligations (99 In this area, one is dealing, not so much with practice properly so-called, but wit assurances in principle from the Minister for the Environment to the *Folketin* Environment and Planning Committee that installations of this type are generall not acceptable in Ramsar sites (100). As a starting point, however, it is clear th the regional authorities are bound to respect such assurances, and so is th Minister for the Environment himself, as the authorizing authority in relation to th regional plans.

5.6.4. Wind Turbines

The application of regional plans to wind turbines is described in Chapter 5.5. (and note 73). Furthermore, there is a detailed review of the Environment Appeal Board's practice concerning wind turbines in Ellen Margrethe Basse: th Environmental Appeal Board (in Danish — English summary available) (Ga 1987) pp. 375ff. (101), drawing the conclusion (cf. p. 378) in relation to conserva tion interests that, to justify a refusal under the Urban and Rural Zones Act, th Appeal Board "as a general rule will require that the area be designated a particularly worthy of conservation (i.e. normally e.g. Ramsar sites) and that ther are concrete grounds for considering the establishment of a small wind turbine be in decisive conflict with *important national conservation interests*" (author emphasis) (102). In review of practice, however, no distinction is made betwee large and small wind turbines and between the establishment of wind farms (turbines and more) and individual turbines (up to 2 wind turbines).

There are four important rulings from the Environmental Appeal Board whic concern Ramsar sites. In rulings of May 17, 1982 and June 17, 1987 one wir turbine (Siø) and 3 wind turbines of 99 kW (Billum) respectively, were permitte while in rulings of July 31, 1986 and February 23, 1988 permission was refuse for 5 wind turbines (Aggersborg) (103) and 3 turbines of 95 kW (Binderup), re spectively.

The reasons stated for the two refusals were that the Appeal Board "on the bas of overall consideration is in agreement that the aforementioned area has partic larly important landscape qualities. In this evaluation the Appeal Board *ha attached importance to the status of the area* (author's emphasis) as a Rams site and EC Bird Protection Area". In the 1986 case, the National Forest an Nature Agency replied to the County Board's enquiry as to whether the installatic would be in conflict with the Ramsar Convention, by saying that wind farms in Ramsar site, regardless of the lack of documentation for specific detriment effects (the applicant had argued that no detrimental effects were documented would "mean a deterioration in the relevant area's function as an all-year wate fowl habitat". In the second case of a refusal (1988) the Agency had referred th Appeal Board to the fact that there had previously been opposition to the esta lishment of wind farms in the relevant Ramsar site (103a). The grounds quoted b the Environmental Appeal Board are based on actual assessment of the site and its status. The latter is acceptable, while the former is not. Despite the Appeal Board's relatively independent position (104), it is of course also bound by the international obligations undertaken by the Government. The Appeal Board must therefore not evaluate the area independently. It is perhaps a little more doubtful whether the Board possesses any authority to assess the installation in relation to international obligations. In any case, the Board would be on thin ice if the assessment of these questions by expert bodies, in this case the National Forest and Nature Agency, were to be contested. In fact the Board has no powers of supervision over the Convention, its interpretation or administrative practice on a national level. It appears that a second Board — the Chief Conservation Board — has had a far better understanding of this problem, cf. below in Chapters 5.6.6., 5.7. and 5.8. (105).

The questionable nature of the Appeal Board's practice is further emphasized by the fact that — to put it rather bluntly — by its 1987 permit the Board in principle allows considerations of energy policy to take precedence over broad conservation interests, even though those energy policy considerations were also present in the 1988 decision, where they yielded to similar conservation interests!

On the other hand, the Environmental Appeal Board, for various reasons and perhaps without fully appreciating the legal position, appears in practice by and large to have respected the Convention's requirements in other respects (cf. Chapter 5.6.5).

5.6.5. Other Decisions
In conclusion, in this section a number of individual decisions, of the Environmental Appeal Board, as well as of the central administration, will be described.

5.6.5.1. Environmental Appeal Board Rulings
In the Environmental Appeal Board ruling of October 20, 1986 a decision of the National Agency of Environmental Protection pursuant to the Marine Environment Act concerning the dumping of a relatively modest quantity of sand fill with a low pollution level in a "fairly closed" Ramsar site (Krik Vig) was upheld (cf. Chapter 5.3.2.).

The National Agency of Environmental Protection's decision was based on the reservations of the National Agency for Protection of Nature, Monuments and Sites. That Agency's statement to the Appeal Board declared that, due to "the obligations undertaken by society, nationally and internationally", only very limited intervention could be accepted. The Appeal Board stated that no information had been adduced to justify overturning the National Agency of Environmental Protection's ruling "when the biological interests related to the site are considered".

However, in a second ruling of September 21, 1987 the Appeal Board upheld a dumping permit in a second Ramsar site, namely the Wadden Sea, in respect of dredge spoil from Esbjerg Harbour (106).

The material dumped in this case was, however, not assessed as containing heavy metals and dumping was to take place in an area of shifting materials. Furthermore, it appears from the ruling that the additional costs of dumping in an area which the National Forest and Nature Agency had indicated "would not be decisive if (this site) had to be preferred for environmental reasons". This means that if there had been environmental risks the result, even though it gave rise to additional costs, would presumably have been different.

In its ruling of December 1, 1982 the Environmental Appeal Board upheld a refusal under the Rural and Urban Zones Act concerning the establishment of a landing site for tourist sightseeing flights. Part of the approach route was to lie over a Ramsar site (107). In its decision the Appeal Board "placed special emphasis on the fact that a section of the area to be used fell under the Ramsar Convention, due to its great importance for birdlife".

In this case the National Agency for Protection of Nature, Monuments and Sites had pronounced that the establishment of a sightseeing flight route was incompatible with the obligations under the Convention. In this connection the Agency referred to a negative decision of the Chief Conservation Board of July 1, 1980 in a similar case, where the Agency stated that sightseeing flights would "be in conflict with ... the conservation interests on which the Ramsar Convention is based". (A second case, in 1978, where the grant of an Urban and Rural Zones Act permit for sightseeing flights from property situated in a Ramsar site was reversed, is described below in the Environmental Appeal Board's decision concerning a shooting range.)

In a decision of December 4, 1981 the Environmental Appeal Board reversed a County Board's grant of a permit for a shooting range in a Ramsar site. In the decision it is stated that the Appeal Board "finds it unfortunate to locate a shooting range inside the boundaries of a Ramsar site".

In the decision consideration of the Ramsar site is accorded less weight than "the considerable deterioration and limitation of ... the area's recreational value" and the presence of an adjacent wildlife reserve. It is also secondary that the Agency "attaches great importance to the area being designated in the regional plan as an area of special conservation interest". (Compare with Chapter 5.6.4.)

5.6.5.2. Other Rulings, etc.

In a letter dated November 28, 1980 the National Agency for Protection of Nature, Monuments and Sites objected to a local planning proposal which allocated part of a Ramsar site for retail trading and housing purposes. The case is virtually the only one in which the reduction of a Ramsar site has been considered on the basis of inappropriate delineation at the time of the designation, and against compensatory measures for improved conservation of the remaining section of the site, cf. the revocation of the objection in letter of August 26, 1982. As the plans for the establishment of a supermarket in the area were abandoned, however, the Town Council cancelled the local planning proposal in 1983.

In a letter of December 7, 1981 the National Agency for Protection of Nature, Monuments and Sites put forward an objection to a local plan for a short-wave transmitter at Buksør Odde. The area (at that time) was in formal terms neither a Ramsar site nor an EC Bird Protection Area, but one of the grounds for the objection was the importance of the area for waterfowl in an *international context* (author's emphasis) (compare with Chapter 3.5. and the general obligations of the Ramsar Convention, cf. Chapter 2.3.).

In a decision of September 13, 1982 the National Agency of Environmental Protection confirmed a county council's permit for the establishment of a fly ash dump close to a coastal area, where complaints had been made about a provision for the recirculation of ash rinsing water. The decision states that it is important that the level of pollutants in discharges to coastal waters does not significantly exceed the natural level for those substances and that "a more stringent attempt must be made to comply with (this principle) in e.g. Ramsar sites".

Conflicts concerning *electricity and road installations* seldom occur, presumably because the conservation authorities (County Council/Greater Copenhagen Council in respect of small installations and the National Forest and Nature Agency in respect of large installations) become involved in planning at an early stage, by virtue of being the permitting authority under the executive orders issued in accordance with the Conservation of Nature Act (nos. 612 and 613 of December 1, 1978). In other cases, potential problems are settled as an element of regional planning. Furthermore, according to the rules, far-reaching examination of natural resource impacts must be carried out. To a considerable extent the maps upon which administrative action is based include Ramsar sites among the areas of conservation interest which are given the highest priority.

5.6.6. The Conservation of Nature Act's Building and Protection Lines: the Practice of the Chief Conservation Board

Chapter VI of the Conservation of Nature Act dealing with building and protection lines has naturally been an important instrument in the conservation of wetlands, including the implementation of the Ramsar Convention. In fact, Chapter VI concerns wetlands to a great extent. Sections 47a and 46, respectively, refer to conservation zones of 150 m around lakes (3 hectares and more), as well as along public watercourses (with a bottom width of 2 m and more) and of 100 m along beaches. In line with the other provisions of Chapter VI, these restrictions do not give rise to compensation. The provision concerning beach protection zones is particularly important for Ramsar sites, as it broadly prohibits a number of activities (building, cultivation, fencing, etc.) in this zone, which stretches from the beach to 100 m from the line at which terrestrial vegetation commences. The importance of the provision in this context derives primarily from the fact that the Ramsar sites include large coastal areas. These provisions are administered by the conservation boards, with the Chief Conservation Board as the appeal authority.

The Chief Conservation Board's practice in cases concerning protection lines under the Conservation of Nature Act (particularly Section 46) is extremely strict (see Chapter 6.2.) where areas within the Ramsar Convention (and the EC Bird Protection Directive) are concerned. This appears expressly from the Chief Conservation Board's decision of October 31, 1984 affirming the Conservation Board's refusal to permit the establishment of a salt water fish farm, and is implicit in the decision of November 26, 1980 where a permit granted by the Conservation Board for the damming of an area in a Ramsar site was unanimously overturned, exclusively on the grounds of the obligations under the Ramsar Convention.

Similar decisions are described in notes 90, 100 and 105 and in the decision of May 2, 1986 with affirming of the Conservation Board's refusal of a permit for a high-water dike. In this connection, the similar practice in cases involving dispensation from conservation provisions can also be mentioned, see the decision mentioned in Chapter 5.6.5.1. concerning sightseeing activities.

In a decision of June 2, 1988 the Chief Conservation Board did accept the establishment of a duck farm in a Ramsar site, but it appears from the decision that, in principle, the view of the Chief Conservation Board was that permit decisions for activities in beach protection zones inside Ramsar sites or EC Bird Protection Areas must be made on the assumption that the Ministry of the Environment does not oppose the activity as being incompatible with the Ramsar or EC obligations, respectively (and this was not so in this case) — see Chapter 5.6.4. Furthermore it appears that even the approval of measures to improve nature conservation in such areas will normally be subject to a condition that such measures form an element of an overall nature restoration plan (107a).

So far (to the best of the author's knowledge) there are no decisions where the Chief Conservation Board has disregarded the Ramsar Convention. Cf. also Chapter 6.2. below.

The Kittiwake (Rissa tridaetyla) is a species of gull which normally only breeds on cliffs. In the Nordre Rønne group of islands (Ramsar site no. 9) near Læsø there is a small breeding colony.

6. Conservation, etc.

6.1. Introductory Remarks

From Chapters 2.6. and 4.3. it appears that only a small number of the 26 designated sites were protected by conservation or similar orders at the time of their notification as Ramsar sites.

Although not included expressly in the Convention obligations, as mentioned in Chapters 4.3. and 4.6.1., fulfilling the objective of the Convention implies an effort to gather the notified areas under a permanent, distinctive conservation scheme (108). Since designation, the number of sites subject to direct conservation has also been considerably increased through conservation orders under Chapter III of the Conservation of Nature Act (conservation regulations) and Section 60 of the same Act (ministerial conservation orders), (108a).

Conservation orders predominantly concern territorial waters, which form the greater part of the sites originally designated (almost 5,000 km^2 of a total of approx. 6,000 km^2). After the notification of the Wadden Sea, etc., as site no. 27 in 1987 relative sizes have shifted slightly (a little over 5,950 km^2 of a total of almost 7,400 km^2).

In 1980 the area subject to certain restrictions under the Hunting and Game Management Act amounted to approx. 230 km^2, corresponding to a little under 4 pct., while in 1987 it had grown to approx. 640 km^2, i.e. a little over 10 pct. (with the exception of the Wadden Sea).

In the same period wildlife reserves (under the Hunting and Game Management Act), of which some are subject to conservation according to the Conservation of Nature Act, have increased from almost 120 km^2 to approx. 890 km^2, although the Wadden Sea Wildlife Reserve accounts for almost 750 km^2. In 1978 38 out of a total of 75 wildlife reserves were situated in Ramsar sites. On January 1, 1988 the figures were 40 out of 82 (108b). Wildlife reserves, however, do not ensure protection of the biotope but usually only regulate or prohibit hunting or public access. Controls of this sort are nonetheless also important in relation to the objectives of the Ramsar Convention.

In 1987, the Wadden Sea, with certain adjacent coastal areas, became a Ramsar site, totalling approx. 1,400 km^2, of which a little over 1,000 km^2 are subject to conservation. The total area of all the sites thus now totals about 7,400 km^2, of which approx. 1,640 km^2, (a little more than 20 pct.), are subject to conservation. In addition there are certain areas subject to conservation under the Sand Drift Act.

Furthermore, during the same period the government has purchased some areas of land situated in Ramsar sites through the National Forest and Nature Agency, under the provisions of Act No. 230 of June 7, 1972 on the acquisition of real property for recreational purposes, etc. This involves a little over 10 km^2 in total (108c).

All conserved (and state-purchased) areas are naturally protected by conservation provisions against all forms of destructive or disturbing activities, just as the general public's access to and use of the areas will usually be subject to regulation. In conserved areas in territorial waters, the extraction of raw materials and drilling activities are usually also prohibited or closely regulated, but on the other hand regulatory measures directed at commercial fishing have been taken on very few occasions, with the possible exception of digging for cockles (108d). This, however, does not prevent occasional problems, sometimes with political overtones or undertones, while the principles of the Ramsar Convention can be of significance in the operation of the provisions for derogations which frequently exist.

6.2. Conservation Regulations

A large number of conservation regulations have been issued for Ramsar sites in the form of Board decisions under Chapter III of the Conservation of Nature Act. In these cases, the Ramsar Convention has always played a large role in the Chief Conservation Board's arguments for imposing conservation regulations, where to do so was possibly inconsistent with an existing general protection by-law (109) or contrary to the regional plan (110), as well as in formulating conservation provisions (111). It has furthermore played a role in achieving conservation and its development in detail in that on the designation of the 26 Danish Ramsar sites in 1977 the Wadden Sea was designated as a future Ramsar site (cf. Chapter 4.1.), regardless of the fact that at the time of its protection it had not yet been designated as such (112).

In conjunction with what has been stated above under Chapter 5.6.6., the conclusion concerning the Chief Conservation Board in relation to the Ramsar Convention must be that no other administrative organ (albeit one which to some extent resembles a court) exists which has, on a similar scale, respected the scope of the commitments of the Convention. In this connection the composition of the Chief Conservation Board gives food for thought (113).

6.3. Section 60 Ministerial Conservation Orders

A number of instances of conservation via executive (ministerial) orders according to Section 60, Conservation of Nature Act have been effected to safeguard Ramsar sites, predominantly maritime areas. Among the most significant are Executive Order No. 166 of April 12, 1984 on Stavns Fiord and Executive Order No. 390 of June 27, 1986 on the conservation of Ringkøbing Fiord. However, the Wadden Sea was made subject to conservation before the notification, cf. Executive Order No. 382 of July 15, 1985, but in the introduction to the Executive Order reference is made to the Government declaration concerning the impending designation of the Wadden Sea as a Ramsar site, cf. also Circular No. 75 of July 15, 1985, Section 2 (cf. also notes 32 and 65). An example of conservation of a state-owned area, situated in a Ramsar site and at the same time subject to conservation according to Section 60, Conservation of Nature Act, is Executive Order No. 222 of March 16, 1984 concerning Harboøre and Agger Tanger.

The Mute Swan (Cygnus olor), Denmark's national bird and a characteristic bird of the wetlands. It occurs in great numbers as a breeding bird and moults its flight feathers in late summer and during the winter. Around half of the European stock is reliant on the Danish wetlands and eight of the Ramsar sites are of international importance for the species.

'. Concluding Remarks

'his review has shown first that the Ramsar Convention has been integrated in irtually all relevant administration and planning (114), albeit somewhat late in ertain respects (Chapter 5.3.5.). Secondly, in virtually all areas, although with a ertain lack of insight and consistency in some respects (Chapter 5.6.4. and ontrary to this Chapters 5.6.6. and 6.2.), implementation accords with the obliga-ons and intentions of the Convention, including at the decentralized level (e.g. hapters 5.5.2. and 5.6.1.) and by Appeal Boards (114a). Thirdly, development as taken place to improve the legislative basis for the administration of the onvention and the fulfilment of its objectives (Chapter 5.3.). Fourthly this evelopment is continuing and can be described as a dynamic process.

n this basis, considering the size of the Ramsar sites (Chapter 6.1.), it is approp-ate to describe Danish ratification of the Ramsar Convention as one of the most gnificant nature conservation measures ever accomplished in this country. How-ver, we can only guess whether this was plain to the Government and the dministration at the time of the ratification. There was probably some degree of onfidence in the strength of legislation, including the efficiency of the planning w reform. Yet experience of the nature and scope of this task was very limited. erhaps this fact partly explains the inherently obvious choice of a national plan-ng directive apparently not even having been considered as an instrument even om the viewpoint of the incumbent duty (115).

arious aspects of constitutional and international law have been referred to hapters 3.5. and 4.5.3. — 4.5.5.), including the possibility of amplifying or pplementing the existing theories on the implementation of treaties (Chapters 1.2., 5.2.) when a planning-related treaty is involved (Chapter 5.5.3.). However, this respect it is not suggested that the conclusions are final or beyond argu-ent (116) (117).

"Vejlerne" in Northwestern Jutland (Ramsar site no. 6) comprise a mosaic meadows, reed marshes, shallow fiords and adjacent agricultural areas.

3. Notes

1) **A. General Literature**

Legal or administrative literature. The Ramsar Convention is apparently not mentioned in Dansk Miljøret ("Danish Environmental Law"), edited by W.E. von Eyben Akademisk Forlag, 1978), nor in the same author's Miljøret ("Environmental Law") (Akademisk Forlag, 1980), although a short account of the Convention is given in the same editor's Miljørettens Grundbog ("A Textbook of Environmental Law") (Akademisk Forlag, 1986) p. 100. A reasonable, although brief, presentation of the Convention and its significance can be found in Ellen Margrethe Basse: Erhvervsmiljøret ("Industrial Environment Law") (Gad, 1987), pp. 138–39. See also Veit Koester, Danmarks Natur (Politikens Forlag, 1981) (in Danish), volume 10, p. 405ff., and Veit Koester: Conservation Legislation and General Protection of Biotopes in an International Perspective, in Environmental Policy and Law, 1984 (ISSN 0378–777 X) p. 106ff., and in European Environmental Yearbook, 1987 (Docter, Milan, Italy) p. 212ff. See furthermore Henrik Knuth-Winterfeldt: Naturfredning i Danmark ("Nature Conservation in Denmark") (DN's Forlag, 1984). Finally, for the conferences of the parties in 1980 and 1984, Danish reports were prepared on the Ramsar sites and the implementation of the Convention in Denmark, namely Poul Hald-Mortensen: Ramsar Convention — Danish Report 1980 and Danish Report 1984, both published by the National Agency for Protection of Nature, Monuments and Sites. The reports contain various information of both administrative and biological character.

There is no overall *conservation description* of the Danish Ramsar sites as such, but most of the sites or parts thereof are described or otherwise analyzed in natural science and topographical literature, etc. Ornithological literature is probably the most comprehensive in this respect.

B. Literature in English, etc. on Danish Ramsar Sites, Legislation, etc.

In *foreign literature* information can be found on some of the Danish and other Ramsar sites in The 1985 United Nations List of National Parks and Protected Areas (ISBN 2–88032–803–9), in E. Carp: Directory of Western Palearctic Wetlands (IUCN, Gland, Switzerland, 1980) and Directory of Wetlands of International Importance (IUCN Monitoring Center, Cambridge, 1987). See also Important Bird Areas in Europe. ICBP Technical Publication No. 9 (IWRB 1989) p. 109ff. (Faroe Islands p. 137 and Greenland p. 145). In this connection, see also Veit Koester (*1984* and *1987*) and Poul Hald-Mortensen (*1980* and *1984*) mentioned in Chapter A. above. The last mentioned publications are also reproduced or referred to in the official reports from the relevant conferences of the parties, see reference in note 7 (the 1980 Conference), Proceedings of the Second Conference of the Contracting Parties, Groningen, Netherlands, 1984 (IUCN, Gland, Switzerland, 1984) and p. 431ff. in Proceedings of the Third Meeting of the Conference of the Contracting Parties, Regina, Canada, 1987 (Ramsar Convention Bureau, Gland,

Switzerland, 1988). In Veit Koester (*1984* and *1987*), the most important biotop
protection provisions are also reviewed, e.g. conservation measures (cf. Chapte
6), the Section 43 system (compare with note 61) and building and protectio
lines (cf. Chapter 5.6.6.).

Concerning *nature conservation in Denmark generally* see H.S. Møller in Natur
schutz anderswo: Natur- und Landschaftschutz in Dänemark (Umweltschutz
Österreichischen Gesellschaft für Natur- und Umweltschutz) 1981/no. 6, p. 1
and no. 7, p. 25ff. and in Danish Wetland Bird Populations and their Protectio
(Ornis Fernica, Supplement no. 3, 1983) p. 104ff. General information on natur
conservation in Denmark can also be found in Jørgen Primdahl: Agriculture
Wildlife and Landscape in Denmark (Institute for Town and Country Planning, Th
Royal Veterinary and Agricultural University, Denmark, 1985) and in Europea
Environmental Yearbook 1987 (Docter, Milan, Italy). See also National Strategie
for Protection of Flora, Fauna and their Habitats (Environmental Series 2, UN
New York, 1988) and Veit Koester: Nordic Countries' Legislation on the Environ
ment (IUCN Environmental Policy and Law Paper, Gland, Switzerland, 1980).
summary of data on nature conservation in Denmark can be found in Manage
ment of Europe's Natural Heritage (Council of Europe, 1987), while the Danis
administrative structure is described in David Baldock et al.: The Organisation c
Nature Conservation in Selected EC Countries (Institute for European Environ
mental Policy, London, 1987). For Danish forestry policy and legislation, includin
the Bill for a new Forest Act (note 117), see Peter Munk Plum and Birgit Honoré i
Environmental Policy and Law, 1988 (ISSN 0378–777 X) p. 111ff.

Most of the *Planning Acts* referred to, particularly in Chapter 5., are considered i
their respective contexts in Ole Christiansen: Comprehensive Physical Plannin
in Denmark (Planning Law in Western Europe, Elsevir, North-Holland, 1986) pp
70–102. Here a general description can be found of national, regional and munic
pal planning. Furthermore a short account of the Danish Constitution and th
administrative system is given. A brief description of Danish environmental an
planning legislation can be found in: Environmental Policies in East and Wes
(Taylor Graham, London, 1987) pp. 100–115 (by Klaus Illum). See also note 71b.

A very thorough analysis of a number of administrative problems, etc., can b
found in Anne Jensen: Comparative Methods for Conflict Resolution in Moder
River and Wetland Management; Volume I, Danish Examples (the Wildlif
Administration of the Danish Ministry of Agriculture, 1988). The study is planne
also to include examples from Australia, Canada, England and the USA.

A number of the *Acts* mentioned have been translated into English. See Jen
Søndergaard: Bibliography of Danish Law (Juristforbundets Forlag, 1973, 198
1985 and 1986).

As to the *Danish literature on international law* to which reference is made, ther
is a Summary in English on p. 515ff. of Ole Espersen: Indgåelse og opfyldelse a
traktater (Conclusion and Implementation of Treaties), cf. note 42, etc. See als

Claus Gulmann: The Position of International Law within the Danish Legal Order, and The Effect of Treaties in Domestic Law, cf. note 46.

The *Greenland Ramsar sites*, cf. Chapter 2.6., are described in the English folder: International Wetlands in Greenland — Ramsar Sites (The Greenland Home Rule, Nuuk, Greenland, 1988).

C. Foreign Literature on the Ramsar Convention

The most thorough review of the *antecedents and contents of the Ramsar Convention,* etc., from legal/administrative as well as ecological viewpoints, can be found in Simon Lyster: International Wildlife Law (Grotius, UK, 1985). Certain legal aspects of the Convention in a broader context are discussed by Françoise Burhenne-Guilmin, et al, in Legal Implementation of the WCS (Environmental Policy and Law — ISSN 0378–777 X, 1986, p. 189ff.). Reference is also made to the literature mentioned in the notes below, as well as the brochure published by the Ramsar Convention's Bureau. In addition, in 1988 the Ramsar Convention Bureau commenced publication of a quarterly newsletter on the Convention.

Comprehensive foreign literature exists on wetlands and their significance. Only a very small amount of this literature is quoted. A very large proportion of this literature derives from IUCN (International Union for Conservation of Nature and Natural Resources (also called the World Conservation Union)) which has a special programme for the conservation of wetlands and thus in many ways provides the biological and natural science input to the Ramsar Convention. The Union, of which Denmark (together with most West European countries) is a member, has more than 600 members, including approx. 60 states and rather more than 120 government organizations, while the remainder of the membership is made up of other national and international institutions and organizations. The Union is domiciled in Gland, Switzerland. For the Union's literature on wetlands, reference is made to its list of publications.

2) Cf. e.g. Edward Maltby: Waterlogged Wealth (Earthscan, UK, 1986).

3) See, among others, Rolf Geckler: Hvad indad tabes — Hedeselskabets virksomhed, magt og position ("What Has Been Lost — the Activities, Power and Position of the Danish Land Development Service") (Gyldendal, 1982), including e.g. p. 100ff. on the Land Development Service's plan to recover 1,500 km^2 of agricultural land by draining 93 low-water fiords and inlets. See also the discussion in the report of the Nature Conservation Commission, Report 1967/467 p. 48ff. and the Open Air Council's open letter to the Danish Government, p. 386.

4) In the explanatory notes to Bill for Act no. 219 of May 24, 1978 on amendment of the Conservation of Nature Act, Official Report of the *Folketing* Proceedings 1977/78, Addendum A, column 2623, an example is given to the effect that (at that time) only 20–25 pct. of the wetlands which existed at the beginning of the century were still in existence. Details can also be found in: Status over den

danske plante- og dyreverden ("Status of the Danish Flora and Fauna") (Nationa Agency for Protection of Nature, Monuments and Sites, 1980 and 1982) p. 167ff and p. 249ff. Other examples are cf. Cyrille de Klemm in Proceedings from the Second Conference of the Parties to the Convention (see note 1 under foreign literature) p. 221, which describes an 85–90 pct. reduction of wetlands in Switzerland since 1800 and of 33–50 pct. in the USA. Dwayne R. J. Moore, etc. in Conservation of Wetlands (Biological Conservation 47,1989, p. 203ff), notes a loss of more than 50% of original wetland areas in the USA and Southern Canada. On the global scale, half of all wetlands are estimated to have disappeared since 1900 (Cheryl L. Jamieson in Pace Environmental Law Review, Vol 4, no. 1/1986 p. 179).

(5) Report 1967/467 pp. 276 and 285 (cf. note 3).

(6) Simon Lyster op. cit. (note 1, Chapter C) p. 183.

(7) Proceedings of the First Conference of the Parties to the Convention in 1980 (Istituto Nazionale Di Biologica, Volume VIII, 1982) p. 69f. The question of "wise use" naturally holds a special dimension in a development context. The concept was therefore also discussed at the Third Conference of the Parties in 1987, wher a recommendation in this respect (3.3.) with special guidelines was adopted (p.119 in the Conference Proceedings — see note 1, B) and further initiatives towards a definition of the concept were resolved on, cf. p. 339ff in the Conference Proceedings (see note 1, B), IUCN Bulletin Vol. 18, Nos. 7–9 (Gland, Switzerland, 1987) pp. 5–6 and 12 and Environmental Policy and Law 1987 (ISSN 0378–777 X) p. 179ff. and p. 201ff. For the Ramsar Convention, as for othe conventions, continual development naturally takes place through the adoption a conferences of the parties of recommendations and resolutions which explain amplify or supplement the provisions of the Convention. Where such documents are respected they may gradually achieve the character of customary law, since they can be assumed to be upheld by national or international courts. See also Françoise Burhenne-Guilmin, op. cit. (note 1) p. 205.

(8) Cf. i.a. Simon Lyster, op. cit. (note 1) p. 187ff. and Conference Report, op cit (note 1) recommendation C.3.1. (p.119) with guidelines in Annex (p.130). A more complete revision of criteria for identifying wetlands of international importance is expected to be agreed upon at the forthcoming fourth meeting of the Conference of the Contracting Parties in Switzerland (1990). See Notification 1989/5 of the Ramsar Bureau (March 31, 1989).

(9) Simon Lyster, op. cit. (note 1) p. 187.

(9a) Concerning the monitoring of Ramsar sites (and EC bird protection areas reference can be made to Naturovervågningen i Danmark ("Nature Monitoring in Denmark") (The National and Forest and Nature Agency) p. 6 and p. 18, where systematic monitoring of these sites, which commenced in 1987, is based on international obligations. Cf. also Overvåning af EF-Fuglebeskyttelsesområde

987 and Overvågning af EF-Fuglebeskyttelsesområder 1987–1988 ("Monitoring EC Bird Protection Areas") (The National Forest and Nature Agency and the ildlife Administration of the Ministry of Agriculture, 1988 and 1989) and Peder gger and Claus Helweg Ovesen: Monitoring Wildlife — an Example of Pro-amme Setup in Denmark (Proc. VIII, the International Symposium on Problems Landscape Ecological Research, October 1988. Int. Ass. Landscape Ecology ALE). East-European Reg. Secr.). See note 26. The monitoring of the wetlands defined by the Convention without simultaneous designation to the List of etlands of International Importance is partly determined, cf. Chapter 2.2., by gional supervision further to the Conservation of Nature Act, as well as nature onitoring as such, which also implies monitoring of certain nature categories, cf. e above publication on nature monitoring in Denmark, and Naturen i Danmark - status og udviklingstendenser ("Nature in Denmark — Status and Develop-ent Trends") (The National Forest and Nature Agency, 1988). General supervi-on under the Conservation of Nature Act does, however, also play a role in lation to the Ramsar sites.

0) There are several instances where governments, federal states or regional uthorities have for political reasons been obliged to abandon "attacks" on wet-nds on the List, at least for the time being, e.g. the Austrian State project for a ydropower plant in the Hainburg Forest on the Danube, cf. item 16, p. 20 of the ouncil of Europe document T-PVS(86) 20 of January 27, 1987. See also Daniel avid in Arctic Heritage — Proceedings of a Symposium (Canadian Universities r Northern Studies, 1986) p. 490. There are also a few court judgements in this spect. An example is a judgement of June 2, 1987 from the "Greek Council of tate" (Supreme Administrative Court) which, on the complaint of a local farmer he Convention was considered to be self-executing so that citizens could invoke ovisions in the Convention vis-à-vis the courts) reversed a government permit to stablish a shipbreaking yard in Nestos Delta, one of the 11 sites designated by reece to the List of Wetlands of International Importance. Greece's ratification as authorized under a special Act to this effect. The then government desig-ated the 11 sites (1974/75), although (to date) there is no official map of the tes, cf. p. 26 (item 85) of Doc. C. 3.6. to the Third Conference of the Parties 987) and p. 55 in Urgent Action Plan to Safeguard Three Endangered Bird oecies in Greece and the EC (Ornis Consult. Report to the Commission of the C, December, 1988). On the latest development concerning detailed information oout the Greek sites see, however, the Ramsar Quarterly Newsletter No. 2, p. 3. hich also contains a summary of the Greek Council of State decision. See also ote 27 below. A second example is the TAR LAZIO judgement no. 1495, Sep-mber 22, 1987, in Tribunali Amministrativi Regionali, 1987. Part I p. 3315ff., hich in particular illustrates the relation between the Italian State and Regions of e country concerning the Ramsar Convention. The judgement upheld *that* the ate has full jurisdiction on matters dealt with by a treaty, even where jurisdiction ver such matters has been transferred to the Regions; *that* a national decree esignating a Ramsar site is therefore sufficient to protect the site against any etrimental activity; *that* this protection is automatic once the designation has een made, and *that* if expropriation is necessary, it must be performed by the

Region which is also responsible for the management of the site (informatio provided by Mr. Cyril de Klemm).

(10a) See Claus Gulmann in "Juristen" 1988 (in Danish), p. 288f.

(11) Daniel Navid, op. cit. (note 10), p. 490f.

(12) The Danish Ministry of Foreign Affairs' Executive Order No. 55 of August 1 1987 on the protocol of amendment to the Ramsar Convention of December 1982.

(13) See Françoise Burhenne-Guilmin, op. cit. (note 1), p. 197. See also Confe ence Report, op. cit. (note 7) "Resolution on Provisional Implementation". Cor cerning the conference see also the Minister for the Environment's reply of July 1 (July 9) 1987 (D 87–832–39) to questions 508 and 510 (General Section/Anne 893) from the *Folketing* Environmental Committee concerning the most importar international agreements, etc. concerning migratory birds and the desire improvement of these agreements. It is stated in the reply that inter alia the thre most important problems are: 1) securing accession to global agreements suc as the Ramsar and Bonn Conventions (22) by as many countries as possible (se Françoise Burhenne-Guilmin, op. cit. (note 1), p. 197; 2) bringing the rich Wester countries to contribute financially to the bureau functions to a sufficient exten and 3) ensuring that all parties actually fulfil the obligations which have bee accepted. The Danish Government has furthermore emphasized in: Our Commo Future. The Danish Government's Action Plan for Environment and Developmer (Ministry of Foreign Affairs, Copenhagen, 1989) on follow-up to the Brundtlan Commission's recommendations and the Environmental Perspective for the Yea 2000 and Beyond (see note 22), that Denmark will contribute to promoting strengthening and expanding cooperation within the framework of the above Cor ventions (p.67). This also applies to the Berne Convention. (See note 24).

(13a) I.M. Smart: International Conventions (IWRB Special Publication No. (IWRB, England, 1987, ISSN 0260 0260–3799) p. 124, which includes a account of the Convention's geographical scope, in global terms, a list of membe states as of approx. January 1, 1987, and the wetlands notified to the internation list, together with their size.

(14) The Results From Stockholm (Erich Schmidt Verlag, Berlin, 1973) p. 1 Some of the principles of the Stockholm declaration are considered in more deta by David H. Ott in Public International Law (Pitsman, UK, 1987) p. 289ff. by B Johnson, op. cit. (note 17) and Louis B. Sohn, op.cit. (note 20). For the Stockhol Conference see also J.C. Starke: Introduction to International Law (Butterworth Ninth Edition) p. 384ff.

(15) The World Charter for Nature (Erich Schmidt Verlag, Berlin, 1986). Concer ing the "legal" nature of the document, see in particular p. 130 (item 0.7.)

(16) Cheryl L. Jamieson, op. cit. (note 4) p. 179.

(17) Bo Johnsen: International Environmental Law (Liber, Stockholm, 1976) p. 30, Douglas M. Johnston (ed): The Environmental Law of the Sea (IUCN, 1981) p. 387, the Brundtland Commission's Report on Environment and Development: Our Common Future (Oxford University Press, 1987) p. 271f. and Niels Madsen in Højesteret 1661–1986 ("The Supreme Court 1661–1986") (Gad, 1986) p. 35, note 7. See also Article 18 of the Vienna Convention of May 23, 1969 on Treaty Law and Ellen Margrethe Basse: Erhvervsmiljøret ("Commercial Environment Law") (Gad, 1987) p. 33ff., particularly p. 34.

(18) In the explanatory notes (to amendment of Section 60) to the Bill put forward on February 18, 1988 for an Act to Amend the Conservation of Nature Act (L No. 213), reference is made to these provisions in the Convention on the Law of the Sea. See also notes 62, 69 and 108a.

(19) The Brundtland Commission's Report on Environment and Development also appears to uphold this general principle, cf. op. cit. (note 17) p. 330. See also Veit Koester: Miljøet og de almaegtige stater ("The Environment and the All-Powerful States") (Politikens Kronik, April 27, 1987) and: From Stockholm to Brundtland (in "Landet og Loven", Ministry of the Environment, 1990).

(20) The World Charter for Nature, op. cit. (note 14) p. 180 (item 214) and Our Common Future (note 17) p. 348ff. The principle also appears from the Ramsar Convention, cf. Article 5, by which the parties must consult with each other, particularly concerning common water systems and wetlands affecting several parties' territories. Provisions of this nature can also be found in the conventions mentioned in Chapter 3.3. See also principle 21 of the Stockholm Declaration (cf. note 14), Louis B. Sohn in Harvard International Law Journal, Vol. 14, 1973, pp. 485; cf. P.C. Mayer-Tasch in Ambio (Stockholm), Vol. XV, 1986, p. 240; David H. Ott, op. cit. (note 14), pp. 291 and 292; Douglas M. Johnston, op. cit. (note 17) p. 46ff.; the World Environment 1972–1982 (UNEP, Nairobi, Kenya), and UNEP's principles on "shared natural resources" (Environmental Law — Guidelines and Principles No. 2, UNEP, Nairobi, Kenya), cf. Françoise Burhenne-Guilmin, op. cit. (note 1, Chapter C) p. 201, W. Riphagen: "The International Concern for the Environment in the Concepts of the Common Heritage of Mankind" and "Shared Natural Resources" p. 343ff. in Michel Bothe: Trends in Environmental Policy and Law (IUCN, 1980) and A.O. Adede in Environmental Policy and Law (ISSN 0378–777X), 1979 p. 66ff. Further reference is made to J.C. Starke, op. cit. (note 14) p. 389, and to Encyclopedia of Public International Law (North-Holland, 1986) p. 122 f, which also mentions the Nordic Convention on the Protection of the Environment of February 19, 1974, which came into force on October 5, 1976, printed in English on p. 47ff. in Cooperation Agreements between the Nordic Countries (Nordic Council, 1978). See also concerning this Convention: Det europæiske Miljøsamarbejde ("The European Environmental Cooperation") (National Agency of Environmental Protection, 1988) p. 58f., Carl August Fleischer, Tidsskrift for Retsvidenskab 1976, p. 83 ff and Alexandre Charles Kiss:

Survey of Current Developments in International Environmental Law (IUCN, 1976) p. 35f.

(21) Cf. the so-called Wadden Sea Declaration of December 9, 1982 (signed by the three countries' Ministers for the Environment) in Minutes, Reports and Declarations from the Third Meeting on the Protection of the Wadden Sea (the National Agency for Protection of Nature, Monuments and Sites, 1983 — ISBN 87–88030–39–3).

(22) Cf. e.g. the Preamble to the Convention of June 23, 1979 on the Conservation of Migratory Species of Wild Animals ("the Bonn Convention"), cf. the Ministry of Foreign Affairs' Executive Order No. 83 of September 15, 1986, according to which "the States are and must be the protectors of the migratory species of wild animals that live within or pass through their national jurisdictional boundaries". The Convention came into force on November 1, 1983 and was ratified by Denmark on June 5, 1982. There are also a large number of bilateral and multilateral agreements on migratory animal species. A list, with a special analysis of these agreements prepared by Cyrille de Klemm, can be found in Migratory Species in International Instruments (IUCN, Gland, Switzerland, 1986). See also Cyrille de Klemm in Environmental Policy and Law 1985 (ISSN 0378–777 X), p. 81ff., and concerning marine mammals P. Birnie in Ambio, Stockholm, 1986, p. 137ff. Under item 3, k of the UN Resolution on "The Environmental Perspective for the Year 2000 and Beyond" (UN/GA Decision No. A/RES/42/186 of December 11, 1987 — see p. 115 in UNEP 1987 Annual Report, Nairobi, 1988) occasioned by the Brundtland Commission's report (cf. note 17) the protection of species is considered to be "a moral obligation of humankind". See also note 24, ruling of December 7, 1981 in Chapter 5.6.5.2.; J.C. Starke, op. cit. (note 14), p. 383; and Alexandre S. Timoschenko (Institute of State and Law, Moscow, USSR): Protection of Wetlands in International Law (contribution to the Wetland Conference at Lyon University, September 1987, IUCN Environmental Policy and Law Occasional Paper No. 4), p. 5, which apparently assumes an obligation under international law to protect our "common heritage", although recognizing that there (as yet) are no international judgements from which this principle can be derived. Indications to this effect are also found in the preamble to the ECE (Economic Commission for Europe) Declaration on the Protection of Flora, Fauna and their Habitats (ECE/ENVW 17/3/ Add. 2 of March 11, 1988), which is included as an Annex to the ECE's environmental strategy on follow-up of the Brundtland Report on a regional level. See "Regional Strategy for Environmental Protection and Rational Use of Natural Resources for the Period up to the Year 2000 and Beyond" (document ENV/R 195 of March 17, 1988). Both the declaration and the strategy are published as separate booklets (UN, New York, 1988). Reference can also be made to Article 3 of the General Principles concerning Natural Resources and Environmental Interferences prepared by the Experts Group on Environmental Law to the World Commission on Environement and Development. See Environmental Protection and Sustainable Development (Graham and Trotman, London, 1987) p. 45ff. and Our Common Future op. cit. (note 17) p. 348.

Finally, reference can be made to the European Court ruling of April 27, 1988 in case 252/85 (the EC Commission versus France) concerning the lack of implementation of the EC Bird Protection Directive. The Court laid down (under item 15) that national legislation for the protection of wild birds on the basis of the concept of "the national biological heritage" (like French legislation) conflicts with the Directive (Art. 1). The court referred to the ruling of July 8, 1987 in case 262/85 (against Italy), from which it appears that "protection of migratory species is a typical trans-boundary problem, for which the member states hold a common responsibility". In the introduction to the ruling (item 5) wild birds are described as "commons".

As an illustration of the international responsibility for migratory animal species, reference can also be made to the Norwegian Minister for the Environment's letter of August 11, 1982 to the Minister's Danish counterpart concerning the detrimental consequences for the Nordic stock of migratory birds of a possible short-wave installation on Buksør Odde in the Liim Fiord (Limfjorden), and conversely the Danish Minister for the Environment's telex of August 29, 1984 to the Norwegian Minister on similar consequences of the possible location of a radio transmitter in Grandefjära, Sør-Trøndelag. Neither of these installations has in fact (so far) been erected. The EC Commission also contacted the Danish Government in 1982 concerning plans for a short-wave transmitter on Buksør Odde, cf. the National Forest and Nature Agency's memorandum of April 21, 1987.

The existence of the Convention Concerning the Protection of the World Cultural and Nature Heritage of November 16, 1972, which came into force on December 17, 1975, in my opinion does not argue against the view that a principle of international law exists which implies certain obligations to protect our "common heritage", particularly concerning habitats for migratory species. In fact, it appears from the World Heritage Convention that the Convention is particularly directed at areas of "*outstanding* universal value" (author's emphasis), i.e. the most outstanding of what can be considered "common heritage". It is furthermore based on a concept of a kind of common responsibility for such areas through the establishment of a special fund to provide financial and administrative assistance for the protection of such areas.

The World Heritage Convention, which together with the Bonn Convention, the Ramsar Convention and the Washington Convention make up the global nature conservation instruments, has been acceded to by more than 100 countries. In addition to cultural monuments proper, however, only about 62 natural areas and 15 mixed cultural and natural areas have so far been accepted for inclusion on the so-called "World Heritage List". Denmark has also acceded to the Convention (see the Ministry of Foreign Affairs' Executive Order No. 21 of 1980), but has not so far nominated any natural or cultural areas to the World Heritage List. The overall Danish-Dutch-German Wadden Sea area has, however, been mentioned as a potential candidate for the list in connection with the aforementioned government cooperation on conservation of the Wadden Sea.

As to the Convention, reference can also be made to Veit Koester *(1981)*, cf. note 1, Chapter A. On the background to Danish reticence in endeavours to include Danish areas on the World Heritage List, see Veit Koester *(1984)*, cf. note 1, Chapter A, and Naturreservater og Feltstationer ("Wildlife Reserves and Field Stations") (National Agency for Protection of Nature, Monuments and Sites 1986) p. 31ff., which also describes UNESCO's Man and the Biosphere (MAB) Programme, which has a certain relevance for both the Ramsar Convention and the World Heritage Convention. For more detail concerning the so-called biosphere reserves, see Man Belongs to the Earth (UNESCO, 1988).

The Convention is exhaustively discussed by Simon Lyster, op. cit. (note 1, Chapter C), p. 208ff.

(22a) Encyclopedia of Public International Law, op. cit. (note 14), p. 125. See also The World Environment 1972–1982, op. cit. (note 20), p. 160f.

(23) Cyrille de Klemm, op. cit. (note 4), p. 235. See the selection of recommendations, etc. (i.a. on different types of wetlands) in: Texts adopted by the Council of Europe in the field of the conservation of European wildlife and natural habitats (Council of Europe, 1989).

(24) See the Ministry of Foreign Affairs' Executive Order No. 84 of September 15 1986 on the Convention of September 19, 1979 on the Conservation of European Wildlife and Natural Habitats ("the Berne Convention" or "the European Nature Conservation Convention"). The Convention came into force on June 1, 1982 and was ratified by Denmark on September 8, 1982, coming into force on January 1, 1983. According to Article 4.1. of the Convention, "appropriate and necessary legislative and administrative measures to ensure the conservation of the habitats of the wild flora and fauna species, especially endangered species" shall be taken, and, according to Article 4.3. the contracting parties undertake "to give special attention to the protection of areas that are of importance for the migratory species". Here the obligations are formulated considerably more strongly than in the Ramsar Convention (cf. Chapter 2.2.). As the sites on the international List of Wetlands will by virtue of their declared international importance plainly also be covered by these obligations under the European Nature Conservation Convention (even though the latter Convention contains no lists) there are several cases in which both sets of obligations have been invoked simultaneously. This also applies in the Austrian case referred to in note 10.

On the other hand, the relatively categorical and yet very broad obligations as to the conservation of biotopes contained in the European Nature Conservation Convention have always been a problem because, perhaps in the absence of provisions for lists or special designations, they have been difficult to convert into practice. See the discussions in the Standing Committee, cf. Report of January 27, 1987, Council of Europe document T-PVS(86)20 p. 9ff. concerning the study from IUCN (Gland, Switzerland, 1986): Implementation of the Berne Convention. See also: the Danish 2-year report on implementation of the Convention in Coun cil of Europe document T-PVS(86)13, revised February 25, 1987; the Standing Committee's decision to try to develop operative guidelines for the interpretation of the relevant provisions, cf. the minutes in Council of Europe document T-PVS(87)40 of January 22, 1988, p. 11ff.; Cyrille de Klemm: An Interpretation of the Provisions Relating to the Conservation of Habitats in the Berne Convention (Council of Europe document T-PVS(88)30 of September 30, 1988) and the final result, namely Resolution 1 (1989) of the Standing Committee on the Provisions relating to the Conservation of Habitats and the three operative recommendations (14–16) in the report of the 8th meeting of the Standing Committee of the Berne Convention (Council of Europe document T-PVS (89) Misc 1).

For domestic implementation of obligations under the European Nature Conservation Convention, see circular letter of February 28, 1986 from the National Agency for Protection of Nature, Monuments and Sites to all counties and the Greater Copenhagen Council, requesting that at least the species appearing on Appendix II of the Convention (strictly protected species of birds, mammals, reptiles and amphibians) enjoy adequate habitat protection through conservation plan guidelines (and that the same should apply to the additional species of birds included in Annex I of the EC Bird Protection Directive by the amendment to the Commission's Directive (85/411 EC) of July 25, 1985). The significance of the European Nature Conservation Convention as distinct from the EC Bird Protection Directive appears from Annex IV of the circular letter mentioned above. This is based partly on an unpublished report (with an appendix) by Erling Krabbe: Bern-Konventionen — en gennemgang af konventionen og forslag til opfølgning af dens bestemmelser ("The Berne Convention — A Review of the Convention and Proposal for Follow-Up of Its Provisions") (National Agency for Protection of Nature, Monuments and Sites, March 1984). This refers to the guidelines and circulars cited in note 83, which in principle give the obligations under the European Nature Conservation Convention equal ranking with the Ramsar and EC obligations.

In principle the same methods are used to fulfil the obligations under the Berne Convention on the protection of habitats as those described (in Chapter 5) in relation to the Ramsar Convention. For example, in a letter of March 7, 1989 from the National Forest and Nature Agency an objection is put forward to a local plan (see Chapter 5.6.5.2) involving construction work in an area which is a habitat for the fire-bellied toad (Bombina bombina). This is a very threatened species which is rare in Western Europe, and included in Appendix II of the Convention.

For the European Nature Conservation Convention, see also Veit Koester *(1980),* op. cit. (note 1, Chapter A); Management of Europe's Natural Heritage (Council of Europe, Strasbourg, 1987), p. 62ff., and Living Nature (World Wildlife Fund) 3/1986 p. 10 f. See further G. Seidenfaden in European Nature Conservation (Council of Europe, Strasbourg, 1984) p. 48ff. and L.A. Batten in IWRB Special Publication No. 7 (IWRB, England, April 1987 — ISSN 0260–3799) p. 118ff., which also reviews the EC Bird Protection Directive.

For other regional agreements, see the register of environmental conventions continually updated by UNEP (United Nations Environment Programme) and to the list in European International Yearbook (Docter, Milan, Italy, 1987) p. 764ff. and in Malcolm Forster: Special Areas in the Sea: Treaty and Legislation Practice (Environmental Policy and Law (ISSN 378–777 X) 1986, p. 179ff.).

(25) The Council Directive of June 27, 1985 on Environmental Impact Assessment (85/337/EC) — see the first sub-report of the Land Committee (Report 1985/1051) p. 7 and p. 76ff. and Act No. 216 of April 15, 1989 amending the National and Regional Planning Acts. The Act does not include territorial waters. Installations in territorial waters, in accordance with Executive Order No. 379 of July 1, 1988, are, however, subject to the same requirements concerning assessment of environmental impacts as installations on land. The Directive came into force on July 3, 1988.

(26) Council Directive (EC) No. 409/79 of April 2, 1979 (in force, April 2, 1981) on the Conservation of Wild Birds (and the Council Resolution of the same date, see note 27) with the Commission's Directive of July 25, 1985 (85/411 EC) concerning amendments to Annex I which is central to the biotope obligations. For the Directive and the Resolution see: EF-fuglebeskyttelsesområder — kortlægning og foreløbig udpegning ("EC Bird Protection Areas — Mapping and Provisional Designation") (National Agency for Protection of Nature, Monuments and Sites, 1983) and the Danish publication: Fredningsplanorientering No. 3 (National Agency for Protection of Nature, Monuments and Sites, 1983). The designated 111 Danish EC Bird Protection Areas, covering a total of approx. 9,500 km^2, also affect all Danish sites notified to the List of Wetlands of International Importance. For most Ramsar sites, the boundaries coincide with the equivalent EC Bird Protection Areas (e.g. nos. 7 and 9). However, some Ramsar sites lie within the boundaries of the relevant EC Bird Protection Areas (e.g. Ramsar site no. 8). On the other hand, some Ramsar sites (e.g. no. 27, the Wadden Sea with adjacent land areas) include several EC Bird Protection Areas, which in their turn also cover some areas (lakes and navigation channels) which are not included in the Ramsar delineation. Finally, there are a number of EC Bird Protection Areas which have nothing to do with the designated Ramsar sites, regardless of the fact that approx. 1,200 km^2 of the around 2,000 km^2 involved here can be characterized as wetlands. This means that virtually all Danish Ramsar sites, now 27, are also subject to the requirements of the EC Directive. However, the nature of EC law falls outside the scope of this report, and the more detailed consequences of the Directive will therefore not be considered further, beyond stating that the main conclusions of this analysis concerning the legal aspects of conservation (see note 71a) are generally also applicable for those EC Bird Protection Areas which are also Ramsar sites (a total of a little over 2,000 km^2). This appears from many of the actual decisions to which reference is made in connection with the Ramsar Convention, just as several of the administrative rules in general give the two categories of area equal ranking. Otherwise, as stated in Chapter 1.2., this conclusion is a more or less logical consequence of the fact that the obligations according to the EC Bird Protection Directive are not in themselves weaker than

the Ramsar Convention's requirements, in fact the contrary would seem to be the case (see review of the Directive in John Temple Lang, Biological Conservation 1982, Applied Science Publishers Ltd., England) p. 11ff. and the incentive to comply with them is stronger, see the statement concerning the role of the EC Commission in Chapter 3.4. (see note 22 above). Thus no provision of the Directive corresponds to Article 2 (5) of the Ramsar Convention on the restriction of a wetland already included in the List due to urgent national interests (see Chapter 2.3). However, the EC Commission, in a ruling of April 21, 1989/XI/004284, concerning a complaint against the building of the fixed link across the Great Belt (through EC Bird Protection Area No. 98) (a construction project which the Commission did not deny would involve destruction of the area), has nonetheless stated that the public works decision did not in itself constitute an infringement of the Directive. The Commission, inter alia, attached importance to 1) *that* intensive debate had taken place in the *Folketing* of the environmental consequences of the construction works, prior to the decision — 2) *that* as much consideration as possible had been made of the environment, 3) *that* in relation to its size Denmark had designated a very large number of bird protection areas and 4) *that* the bird species involved would be able to become established in other nearby areas (compare the "principle of compensation" in Article 4(2) of the Ramsar Convention — see Chapter 2.3). Concerning the parallel treatment in Denmark of Ramsar sites and EC Bird Protection Areas, reference can for example be made to the Minister for the Environment's letter of December 16, 1986 concerning approval of the Regional Plan of the County of Southern Jutland, in which is included a requirement that the guideline for open landscape is supplemented as follows: "Within areas falling under the Ramsar Convention and *the EC Bird Protection Directive* (author's emphasis), further planning measures may not be taken which conflict with ... the objectives and provisions of the Directive". Reference can likewise be made to general statements in the National Agency for Protection of Nature, Monuments and Sites' report of September 10, 1984, the Minister for the Environment's letter of March 26, 1986 to EC Commissioner Clinton Davies and the Ministry of the Environment's letter of March 24, 1988 to the Ministry of Transport.

Several of the Ramsar sites have also been designated as national areas of biological interest, cf. Fredningsplanorientering No. 2: Nationale biologiske interesseområder ("Conservation Planning Briefing No. 2: National Areas of Biological Interest") (National Agency for Protection of Nature, Monuments and Sites, 1983). Conversely, most of the national areas of biological interest are situated within the so-called "large national nature areas", cf. the National Agency for Protection of Nature, Monuments and Sites' circular letter of April 20, 1979, the maps issued in 1979 and 1982 and the more detailed description in: Danmarks større nationale naturområder ("Denmark's Large National Nature Areas") (National Agency for Protection of Nature, Monuments and Sites, 1984), which thus indirectly provides a description of the natural status of some parts of the Ramsar sites.

According to the Council Resolution of April 2, 1979 (No. C 103/6) (see note 26) Member States were not only to designate special EC Bird Protection Areas but were also to inform the EC Commission of the areas which they had designated, or intended to designate, to the List of Wetlands of International Importance.

(27) There is also a certain coincidence between the obligations under the convention mentioned in note 24 and the EC Bird Protection Directive. NGOs therefore often bring alleged interference with important nature areas before the EC Commission as well as the Standing Committee of the Berne Convention. Cf. e.g. L.A. Batten, op. cit. (note 24), pp. 119 and 120 on Duich Moss in Scotland. In the instance mentioned in note 10 concerning Hainburg in Austria, however, obligations under the Ramsar Convention and the Berne Convention were involved.

(28) See also Niels Madsen, op. cit. (note 17), p. 41.

The EC Court on May 1, 1988 delivered a total of 5 judgements on infringement of the EC Bird Protection Directive, against Belgium, Italy, West Germany, the Netherlands and France. None of the judgements directly concern the Directive's biotope obligations under Article 4, however. Nonetheless some of the general considerations are of significance in this respect. Within this framework an argument put forward by *l'Avocat Général* in the first of the cases (247/85 against Belgium) is of special interest. The consideration is that it is only acceptable to leave discharge of the obligations under the Directive to a national or regional administrative authority if the rule of law which is applicable implies a limitation on the exercise of discretionary powers, which fully respects the Directive's provisions ("Une transposition d'une directive dans l'ordre juridique interne ne doit pas être laissée à un organe de l'administration nationale ou régionale, la norme législative applicable ne fournissant pas à son pouvoir discrétionnaire un encadrement qui respecte pleinement les conditions posées par la directive" — quoted from an unpublished analysis of the aforementioned judgements prepared by Yves Lecocq (FACE, Brussels), p. 4 — item A.3.4.). In principle the same would presumably apply in respect of fulfilment of the Convention obligations. Compare in this respect Chapters 4.5.4. and 4.5.5. and notes 26 and 62.

(28a) These areas and animals might be regarded as "common heritage" (notes 20 and 22). This, however, does not imply that these national resources are outside the scope of national sovereignty and hence in common property or under an international authority, corresponding to the concept of "common heritage" in Article 136 of the Law of the Sea, cf. Carl Fleischer: The Concept of Common Heritage p. 321ff. in Michael Bothe: Trends in Environmental Policy and Law (IUCN, 1980).

(29) Recommendation 99.1 (b) in the Results from Stockholm, op. cit. (note 14) p. 67.

(30) Section II, Chapter 1 of the Official Journal of the European Communities No. C 112 of December 20, 1973, p. 1ff.

(31) Official Report of the *Folketing* Proceedings 1976–77, column 11553f.

32) See also Chapter 2.5. The 26 sites are listed and described in more detail in ne Annexes to the Ratification, reproduced in the National Agency for Protection f Nature, Monuments and Sites' publication in 1980 of the Ministry of Foreign ffairs' Executive Order No. 26 of April 4, 1978 (cf. Chapter 4.7.1. and note 1). he government's declaration on the Wadden Sea is also reprinted there. The Vadden Sea was not designated as a Ramsar site (no. 27) before 1987 (National orest and Nature Agency letter of May 14, 1987 to the Ramsar Bureau). The Vadden Sea, with certain adjacent land areas, is furthermore so far the only)anish site to be added to the list since ratification, despite the fact that in the nemorandum to the government (memorandum of March 23, 1977, attached to ne National Agency's memorandum) it is emphasized that the list does not conain an exhaustive account of Danish wetlands of international importance, but nly a number of the most important ones. A contributory reason for the relatively atisfactory Danish list was undoubtedly that the Ministry of Agriculture (repesented by its Wildlife Administration) and the Ministry of the Environment had rtually parallel interests in the case. The government's declaration on the Waden Sea, etc., can be found on p. 10 of the Ministry of Foreign Affairs' Executive)rder on the Ramsar Convention.

oncerning Ramsar site no. 27, the Wadden Sea, see also note 65 and to Chapers 3.1.; 5.3.1. and 6.3. See also p. 165 and p. 169 in Proceedings of the 5th nternational Wadden Sea Symposium (National Forest and Nature Agency, enmark, 1987, ISBN 87–503–7054–5).

33) The Ministry of the Environment's letter of April 19, 1977 to the Ministry of oreign Affairs, and the Ministry of Foreign Affairs' letter of August 23, 1977 to i.a. ne Ministry of the Environment.

34) See note 32. Furthermore, a memorandum from the National Agency for rotection of Nature, Monuments and Sites states *that* "there is agreement beteen the ministries involved that future decisions concerning the utilization or rotection of those areas of territorial waters inside the designated wetlands" — y far the greater part of the almost 6,000 km2 involved — "must take place in the ght of an overall evaluation of all relevant national interests, in which those nterests particularly related to wetlands represent a factor" , *that* "in this evaluaon the chances of allocating compensatory areas under Article 4 of the Convenon should not be decisive", and *that* these "considerations were essential to the eadiness of the Ministry of Public Works (now the Ministry of Transport) to upport Danish ratification, despite its concern in principle at the according of a pecial status to parts of territorial waters, even though this status is not irrevocble". (It does not appear from the memorandum that other ministries or, e.g., the ssociation of County Councils in Denmark had similar reservations as to land reas, even though these are without doubt potentially exposed to far more intenve, definite and varied utilization. This reservation will be shown not to have een of much practical importance, perhaps because it is expressly confined to ases where a conflict exists between conservation and *"other national interests"* uthor's emphasis), as envisaged by Article 2(5) of the Convention — see Chap-

ter 2.3. In the 1980 circular mentioned in Chapter 4.7.3., ecological consideration
are referred to, not as merely one of a number of national interests but as a
"important element".

(35) Letter of May 16, 1977 from the Minister for the Environment to members o
the *Folketing* Environment and Physical Planning Committees, respectively.

(36) Letter of June 21, 1977 from the Minister for the Environment to the *Folke
ing* Physical Planning Committee (re general section — Annex 18).

(37) Letter of August 31, 1977 from the Minister for the Environment to th
Folketing Physical Planning Committee (re general section — Annex 35). Th
conclusion here is not wrong but very weak in a number of respects when it i
stated that the Ramsar status alone implies an *expectation* that conservation an
hunting authorities, on the basis of an *overall weighting* of the relevant interes
will endeavour to exploit their opportunities sanctioned by legislation to prote
ecological interests in the areas *in the same way as* they *hitherto* have don
(author's emphasis). It is naturally easy to be critical after the event. Sever
political/tactical considerations may have played a role, but at all events it woul
immediately appear that the Committee's interests in the case have made th
Minister for the Environment nervous. Close perusal of the reply almost gives th
impression that the Minister for the Environment is not even completely certai
that after the designation (and ratification) the administration will be as strict a
before, not even when conservation authorities are involved whom he himse
holds the authority to instruct. The description of the weighing of interests is fa
more cautious than in the first reply, which repeated the description in th
memorandum to the government almost word for word (cf. note 34). As wi
appear, it emerged in practice, however, that the Minister for the Environme
underestimated his own capabilities and also that there in fact was no basis fo
the Minister for the Environment's very subdued advance evaluation. For exam
ple, as early as June 16, 1977 a circular according to regional planning legislatio
was issued by the Ministry of the Environment to municipal authorities concernin
the planning of summer house areas. The purpose was i.a. to ensure that all fre
areas of land designated as coastal areas should be exempt from summer hous
construction. According to the circular the delineation of these areas should tak
place with i.a. the designated Ramsar sites as the starting point, cf. also Chapte
4.7.2. and note 56 in this respect.

(37a) The complete wording of Section 19, Subsection 1, of the Danish Constit
tion is as follows:

"The King acts on behalf of the Realm in international affairs. Without parliame
tary consent, however, he may not perform any act which increases or diminishe
the area of the Realm, or enter any obligation which for its fulfilment requires th
concurrence of Parliament, or which is otherwise of great significance. The Kin
may furthermore not without parliamentary consent terminate any internation
agreement which has been entered with Parliament's consent."

'his provision is analyzed by Claus Gulmann in The Effect of Treaties in Domestic .aw (Sweet and Maxwell, London, 1987), p. 29ff.

38) See Act no. 297 of June 26, 1975, cf. Consolidated Act No. 520 of October 1, 975, now Consolidated Act No. 530 of October 10, 1984, as amended by Act No. 55 of May 13, 1987. Cf. Also Chapter 55 of the Hunting Act of that time (Act No. 21 of June 3, 1967).

38a) Whenever reference in the present book is made to the "explanatory otes" of a certain Danish law the following is envisaged:

ccording to the Danish system of legislation, a bill is normally divided into two arts:

) The first part consists of the bill itself and covers the provisions proposed to Parliament.

) The second part of the bill contains the "explanatory notes" which discuss the motivation, interpretation and other explanations of the law. This part is divided into two sub-sections:

   a) The first sub-section contains a general introduction explaining the reasoning behind the bill and the legal, administrative and economic implications.

   b) The second sub-section refers to the individualprovisions and explains their background, defines their special notions, and/or makes more precise their aims, etc.

he Parliament will only adopt the first part of the bill (1) with any amendments ecided upon by the Parliament. But, once the provisions of the bill have been ccepted, the definitions, explanations, etc. of the original bill, any explanations iven by the Parliamentary Committee to amendments by the Parliament itself, nd any answers by the Minister to questions from the Parliamentary Committee 'hich arose during the negotiations, will bind the administration. For this reason e material will also be used in ministerial circulars to subordinated bodies about e law. Furthermore, the material might be used in legal proceedings. If, for xample, there is doubt about the content or intention of a certain provision of a w and the court does not consider the wording of the law and the material escribed above to be in clear conflict, the second part of the bill will frequently lay a decisive role in determining the meaning of the law.

39) Official Report of the *Folketing* Proceedings 1974/75, Addendum A, column 552 and the Conservation of Nature Act — amendments in 1975 (the National gency for Protection of Nature, Monuments and Sites, 1977), p. 9.

40) On the other hand, the rule in Section 60 b is described in the circular from 980 referred to in Chapter 4.7.3. where "the consideration of protection of the

designated sites" is stated to be of "significance for the possible entering of th agreements described in Section 60 b". Compare with note 42.

(41) Cf. now the Ministry of the Environment's Executive Order No. 23 of Januar 25, 1984 (with Annex) concerning application of the Washington Convention Concerning the Convention see also Veit Koester *(1981)*, cf. note 1, Chapter A Simon Lyster, op. cit. (note 1, Chapter C), p. 239ff., and Willem Wijnstekers: Th Evolution of CITES (CITES Secretariat, Lausanne, Switzerland, 1989).

(42) The Conventions mentioned in notes 22 and 24; the Bonn Convention on th Conservation of Migratory Species of Wild Animals and the European Natur Conservation Convention, have both been ratified with reference to Section 60 b Subsection 1, however. But it is simultaneously stated that they do not requir special implementation measures, i.e. that they can be immediately complied with cf. Section 19, Subsection 1, of the Danish Constitution (National Agency fo Protection of Nature, Monuments and Sites' memorandum of March 11, 1982, an resumé of April 22, 1982 with Annex to cabinet session). See also report 1973/68 on the proclamation and implementation of treaties pp. 22f. and 53 and Ol Espersen: The Conclusion and Implementation of Treaties (English summary (Juristforbundets Forlag, 1970), i.a. p. 239ff., p. 266ff. and p. 421ff. The question c the extent of the advance consent and the authorization for the fulfilment of treatie in Section 60 b, Subsection 1, cf. Ole Espersen, i.a. p. 309ff., will not be discusse in more detail in this context.

(43) See report 1973/682 op. cit. (note 42), i.a. p. 21f., and Ole Espersen op. ci (note 42) p. 221ff.

(44) See report 1973/682 (note 42) p. 23 and p. 55.

(45) See the report of the Ministry of Justice in Nordisk Tidsskrift for Internationa Ret 1971 p. 75f., to a certain degree contested by Henrik Zahle in Dansk Forfa ningsret 3 ("Danish Constitutional Law 3") (provisional edition from Christian Ejler Forlag, 1987) p. 127f. (see Claus Gulmann, p. 288, note 8 in Juristen 1988 in thi respect), and Ole Espersen, op. cit. (note 42) p. 235f.

(46) The Ministry of Justice, op. cit. (note 45) p. 80ff. and i.a. Claus Gulmann: Th Position of International Law Within the Danish Legal System in Nordisk Tidsskr for International Ret 1983 (Volume 52) p. 45ff., and the same author's contributio p. 29ff., in the book: The Effect of Treaties in Domestic Law (Sweet and Maxwe London, 1987). The rule of presumption must not be too rigidly applied. It may thu not lead to the "law being exhausted of significance. Respect for the rule international law must have the effect of a proof-reading sign, allowing the main ru of the law to retain its most significant area of application", cf. Ole Espersen, op. c (note 42), p. 388, and Niels Madsen, op. cit. (note 17), p. 39. The principles a considered by Henrik Zahle, op. cit. (note 45), p. 115ff., who is critical towards ther

(47) See Henrik Zahle, op. cit. (note 45) p. 130, and concerning intended and unintended breach of treaty, the Ministry of Justice, op. cit. (note 45) p. 80ff.

(48) Ministry of Justice, op. cit. (note 45), p. 70ff., Henrik Zahle, op. cit. (note 45), pp. 18ff. and 130f., and Ole Espersen, op. cit. (note 42) p. 290.

(49) Ole Espersen, op. cit. (note 42) pp. 382 and 385.

(50) Ole Espersen, op. cit. (note 42) p. 391. Whether treaties constitute an element of domestic law is a different question. As to dualism, monism and the so-called "practical monism" (*Ole Espersen*, cf. note 41, *Niels Eilschou Holm* in Nordisk Tidsskrift for International Ret 1981 p. 126f., *Claus Gulmann*, cf. note 46, and *Niels Madsen,* cf. note 17 being the "practical monists") see Henrik Zahle op. cit. (note 45), p. 106ff. The "practical monists" acknowledge that international law also forms a source of national law, that courts and other authorities responsible for the application of the law may apply international law, even without express authority to do so, and that international law can even produce results different to those which would derive from a clear provision of domestic law (Zahle, op. cit., pp. 107–108). Zahle criticizes practical monism, but his own conclusions, p. 128ff., are hardly different in substance. Therefore, these variations, if they exist at all, probably concern methods rather than substance. See also Claus Gulmann's resumé of the current legal situation in Juristen 1988, p. 289.

(50a) Under Section 33, Conservation of Nature Act, conservation orders ultimately depend on endorsement by the *Folketing* of the disbursement of conservation-related compensation, where the State's share of the sum paid in compensation (normally 3/4) is more than DKK 500,000 (approx. USD 70,000). The *Folketing* has, however, almost never refused to fund compensation determined by the Chief Conservation Board.

(51) Landscape classification maps covering the entire country were issued as early as 1972. See Karnov, 10th edition (1983) note 190 p. 1267. As to electricity installations and roads, see Chapter 5.6.5.2. and Karnov 11th edition (1987) p. 1387f. and Circular No. 26 of February 7, 1980, Section 2.

(52) Ole Espersen, op. cit. (note 42), p. 290.

(53) As to the propriety of using discretionary powers to effect compliance with the terms of a treaty, where such a course does not clearly conflict with the purpose of the Act, see Chapter 4.6. and Ole Espersen, op. cit. (note 42) p. 381ff., particularly p. 384.

(54) See discussion in Ole Espersen, op. cit. (note 42), Chapters 28 and 29, particularly p. 415 and p. 419. Furthermore pp. 157, 292, 323 and 421. See also report 1973/682, op. cit. (note 42), p. 26 and p. 49, and Henrik Zahle, op. cit. (note 45) p. 170ff.

(55) Ole Espersen, op. cit. (note 42), p. 229 with note 16.

(56) See also note 37. The actual delineation of the Ramsar sites (and EC Bir
Protection Areas) in the form of maps has occasionally given rise to doubt on th
grounds that the delineation (when not related to specific geographical charac
teristics such as a road) is not always exact or completely recognizable, an
because the maps which are generally available only exist in a relatively smal
scale. Disputes have arisen over hotels and similar developments, such as th
case in 1988 referred to in note 96. In such cases, until wholly accurate map
become available, the delineation must be "interpreted". There is no doubt tha
sole responsibility in this respect rests on the National Forest and Nature Agenc
of the Ministry of the Environment, since the Government has notified the sites t
the List of Wetlands of International Importance and in this connection has con
firmed their delineation. It is also the Government's responsibility to ensure tha
"interpretation" of the delineation does not give rise to a circumvention of th
obligation under Article 2.5 of the Convention to inform the Convention's Burea
of amendments to delineations. A regional authority cannot, therefore, carry ou
an "interpretation" unilaterally, or through the use of planning powers, as suc
interpretations must be approved by the Ministry of the Environment, unless th
"interpretation" proposed by the regional authority clearly appears from the mate
rial submitted and is independently considered — expressly or implicitly. See als
the provisional (internal) memorandum of June 7, 1988 of the National Forest an
Nature Agency.

For the circulars mentioned in Chapter 4.7.2., see Ole Christiansen and Inge
Våben in Juristen 1983 p. 134ff. In connection with the now revoked Nationa
Planning Directive on the reservation of areas for nuclear power stations (Ministr
of the Environment circular of August 6, 1980) a municipality's request for th
relocation of a reserved area was rejected because the proposed relocatio
would affect a Ramsar site, see Minister for the Environment's letter of July 10
1980 and the National Agency for Physical Planning's letter of the same date, cf
Veit Koester in Danmarks Natur (Politikens Forlag, 1981), Volume 10, p. 405f. I
the National Planning Directive (Ministry of the Environment Circular No. (14000
of May 6, 1986) on the location of TV-2 transmitter stations, the Ramsar sites ar
also mentioned (albeit indirectly, by reference to EC Bird Protection Areas). Th
scope of this reference is, however, to express it cautiously, somewhat cryptic.

(57) See note 34 (and note 40).

(58) Ole Espersen, op. cit. (note 42) p. 292. For the application of circulars i
connection with the fulfilment of treaties after the harmonization of legal provi
sions (the "establishment method" or "passive reformulation"), see p. 235f.

(59) See e.g. H.S. Møller: Ramsar-konventionen trådt i kraft ("The Ramsar Con
vention Came into Force") (Fugleværn, DOF, 1978) No. 8, p. 11f., and in Kaskelo
(1978) No. 35 p. 13ff. Furthermore Hans Ole Hansen: Danske Naturområde
("Danish Nature Areas") (Politiken, 1979) p. 258 and Veit Koester in Dansk

raktiserende Arkitekters Tidsskrift 1981, Nos. 11–12, p. 66 and p. 68ff., and in asen i Rummet ("Oasis in Space") (the World Wildlife Fund, 1981) p. 96f. inally, reference can be made to Jørgen Bent Thomsen in Living Nature (the Vorld Wildlife Fund, 1985) No. 4, p. 6. Cf. also note (1).

i0) This proposal, however, receives some support from the legislative history f the Act (Act No. 219 of May 24, 1978), as reference in the explanatory notes to e Bill (the Official Report of the *Folketing* Proceedings 1977–78, Addendum A, olumn 2630) is made to stricter control in wetlands situated within areas of pecial conservation interest, a description applied to the 26 notified Ramsar sites another context, i.e. overriding factors in connection with Chapter IV of the Act oncerning conservation planning (column 2610). According to the wording of the ircular there is, however, hardly any doubt that the central administration has onstrued the statement as a duty arising from the accession to the Convention. ee also Chapter 5.3.1. In this connection it should also be noted that in the xplanatory notes to the Bill to Section 3 of Act No. 355 of May 13, 1987 on the mendment of various environmental and planning acts, reference is made to the ct that EC Bird Protection Areas (see Chapter 3.4.) — and therefore by implica-on Ramsar sites (see note 26) — are overriding factors for the purposes of onservation planning. This in itself is no innovation, see Chapter 5.5.4. On the ther hand it is not clear, why elsewhere in the explanatory notes (the general troduction, Section 2.1. — see the Official Report of the *Folketing* Proceedings, 986–87, Addendum A, column 10139 ff.) it should be inferred that this overriding haracter is peculiar to the conservation sector and "therefore cannot be laid own in a national planning directive". In the author's view, the Ramsar obliga-ons would be (or perhaps now more correctly would have been, see note 115) an xtremely suitable starting point for a national planning directive, by virtue of the readth of the obligations imposed in relation to the allocation of land-use in ertain defined areas spread throughout the country. Nor does the content of the xplanatory notes to the Bill harmonize very well with the National Planning eport 1986 (Ministry of the Environment, 1986) in which "national and interna-onal interests related to ... Ramsar sites" are mentioned as an example of "the tate's approval of the plans of the Counties (or the Greater Copenhagen Coun-l) *by statement of national planning views* (author's emphasis) in national plan-ng reports, by means of national planning directives and other binding or indica-e statements".

i1) For the legislative history of the 1978 Act, see note 60. The 1983 Act, which as not based on a Government Bill, does not mention the Ramsar Convention in e explanatory notes (to the Bill). As to the two amendments, see also Veit oester in Juristen, 1979, p. 135ff. and 1984 p. 141ff., respectively.

he current provisions of the Conservation of Nature Act Sections 43–43 b (cf. ote 62 in this respect) confer general protection on a number of eco-system pes, without giving rise to compensation. The scheme covers: 1) All water-ourses which fall within the conservation scheme by virtue of being designated the regional authorities, and approved by the Minister for the Environment

(covering approx. 28,000 km of open watercourses, the total length of which i
approx. 40,000 km). 2) Lakes in urban areas or urban built-up areas or lake
larger than 500 m^2. 3) Bogs of 5,000 m^2 or more. 4) Salt meadows larger than
hectares. 5) Heathlands of more than 5 hectares. 6) Sites which in aggrega
reach the above area are also protected. This e.g. means that a body of water c
less than 500 m^2 situated in a salt meadow (of over 3 hectares) is protected. Th
same applies e.g. for a small bog on a heath which together with the bog amoun
to more than 5 hectares.

In general, particularly strong reasons must be shown before a permit is given t
make changes (e.g. cultivation), which are incompatible with conservation inte
ests. The scheme is administered by regional conservation authorities (the Cou
ties or the Greater Copenhagen Council) with the National Forest and Natu
Agency as the appeal authority. Local interest organizations also have a right c
appeal. See Veit Koester *(1984),* cf. note 1, Chapter A.

The provisions of Sections 43–43 b, together with those of the Conservation c
Nature Act concerning protection lines (see Chapter 5.6.6.) are the most impo
tant instruments for achieving the broad obligations in the Convention concernin
wetlands in general, cf. Chapter 2.3. Concerning the management of wetland
however, concrete conservation measures under Chapter III of the Act also play
significant role both within and outside the designated Ramsar sites, cf. note
108a and 109. Consideration of wetlands generally is also included incidentally i
both regional and conservation planning, in the same way as consideration c
other small biotopes.

(62) These professions are borne out in practice. See e.g. the National Fore
and Nature Agency's current information publication, § 43-NYT, entry no. 2
(letter of April 21, 1986) on a decision on the recovery of clay (warp) in sa
meadow areas in an EC Bird Protection Area (which is now also a Ramsar si
with which the EC Bird Protection Areas are generally given equal ranking). Th
states that "the National Agency for Protection of Nature, Monuments and Site
can in principle not accept intervention ... which permanently depletes thes
areas or parts thereof". See also the National Forest and Nature Agency's dec
sion of May 30, 1988 where a project which had already been approved but whic
had not been completed by the time that the rules on protection of salt meadow
came into force was in part refused completion, particularly as a Ramsar site (ar
EC Bird Protection Area) was involved. See § 43-NYT, entry no. 66 concernin
the ruling. In the Bill to Amend the Conservation of Nature Act (L. No. 213) p
forward on February 17, 1988, it is i.a. proposed, with specific reference to variou
proposals for an extension and tightening of the rules on the general protection
ecosystem types, that these provisions can "be seen as an element of continuin
improvement in fulfilment of the objectives of, and obligations under," i.a. th
Ramsar Convention (general explanatory notes Section 1.4.). Furthermore, i.a. i
Section 3 of the explanatory notes to the Section 43 amendments a continuatic
of existing practice is suggested. If the proposals are accepted, the reduction i
the size criteria and the inclusion in the conservation scheme of i.a. fresh-wat

meadows which are proposed will naturally involve considerable expansion of the 400–800 km^2 mentioned in the text. See also notes 108a and 117. As to small biotopes, which are not presently covered by the general protection provisions of the Conservation of Nature Act, see Peder Agger and Jesper Brandt: Dynamics of Small Biotopes in Danish Agricultural Landscapes (Landscape Ecology, 1988, Vol. 1, No. 4, p. 227ff.).

(63) The Official Report of the *Folketing* Proceedings 1977–78, Addendum A, column 2881 ff, cf. also Ellen Margrethe Basse: Miljøankenævnet ("The Environmental Appeal Board") (Gad, 1987) p. 442 (and p. 700, note 30). See also note 3 in Karnov, 11th edition, p. 1511ff. Compare with "Affald 2000" ("Waste 2000") (Ministry of the Environment, 1988), according to which, due to the considerable need for waste dumps, restraint will be exercised in attaching crucial importance to circumstances other than "protection of the groundwater and special nature conservation considerations ... e.g. Ramsar sites" (p. 58).

(64) Official Report of the Danish *Folketing* Proceedings 1979–80, Addendum A, column 913 ff. See also note 45 in Karnov, 11th edition p. 1531 and p. 20 in Guidelines No. 3/1981 from the National Agency of Environmental Protection.

(65) Cf. e.g. Section 9, Subsection 1, of the Ministry of the Environment's Executive Order No. 382 of July 15, 1985 on conservation of the Wadden Sea, which (cf. note 32) was notified as a Ramsar site in 1987 (see also Chapter 6.3. below). The relevant provision does not, however, fully live up to "expectations" in Guidelines No. 3/1981 from the National Agency of Environmental Protection concerning a prohibition on the discharge of waste sewage, since pleasure boats are (still) exempt from the prohibition. Concerning the dumping of dredged seabed materials, under Executive Order No. 975 of December 19, 1986, decision-making is delegated to the regional environmental authorities. The highest authority is the Environmental Appeal Board, cf. Chapter 5.6.5.1. A circular concerning the Executive Order is currently being prepared.

(66) See also note 18.

(67) See note 10, p. 1495 in Karnov, 11th edition, where reference is also made to Article 196 of the Law of the Sea Convention, cf. Chapter 3.1. (cf. Douglas M. Johnston (ed): The Environmental Law of the Sea (IUCN, 1981) pp. 78 and 178. It is also hard to understand, particularly in view of the relevant provision in the Law of the Sea Convention, why Section 4 of the Gene Technology Act, concerning the regulation of the fishing-zone, is generally limited to Danish citizens. It furthermore appears from the Danish Government's Plan of Action for the Environment and Development (see Note 13) that the Government will promote legislation implementing the provisions of the Law of the Sea Convention on protection of the marine environment in the broadest sense (flora, fauna and water quality). See Note 108a.

(67a) This Bill has now been passed as Act no. 812 of December 1988 concerning Hedgerows.

(68) The relevant initiatives should be viewed in the light of, inter alia, the reports of March 2, 1987 from the Minister for the Environment to the *Folketing* and its Environment and Planning Committees, both as to an overall strategy in respect of marginal land, and as to nature conservation and the Folketing's treatment thereof, cf. Official Report of the *Folketing* Proceedings 1986–87, column 10146 ff. In addition to monitoring and restoration, the latter report also mentions the introduction of a nature reserve concept which (cf. also the Nature Conservation Bill's remarks in this respect) may be of significance for areas of "international importance where conservation interests weigh most heavily vis-à-vis consideration of, inter alia, recreational activities and exploitation interests" (cf. also notes 18, 62 and 66). In particular concerning restoration in relation to Ramsar sites among others, see also: Retablering af tørlagte søer og fjorde i Danmark ("Restoration of Dried-Up Lakes and Fiords in Denmark") (Ministry of the Environment, 1987) p. 86f., and Marginaljorder og miljøinteresser ("Marginal Land and Environmental Interests") (National Forest and Nature Agency, 1987), p. 179f.

An interesting decision was taken by the *Folketing* on May 5, 1987. According to this decision the *Folketing* requested the Government to prepare a project for restoration of Denmark's largest river, the Skjern Å, dewatering around 10% of the total area of the country. The course of this river was subject to very vigorous regulation around 20 years ago to reclaim agricultural land. Another objective of the restoration is to recreate the self-purification of the river, to improve the quality of the water, in Ringkøbing Fiord (Ramsar Site No. 2). In the Folketing decision this is described as "an absolute precondition for survival of the Ramsar site". According to the project which has been prepared it will cost approx. DKK 110 million (around US$ 15 million) to restore the river. In a memorandum of May 25, 1989 to the *Folketing* Environment and Planning Committee the Minister for the Environment announced that a restoration of the Skjern Å will be carried out over a 10–15 year period in accordance with the prepared two-phase project. The first phase, which will run over 3–4 years, will be used i.a. for area acquisitions. The estimated costs amount to DKK 30–35 million (approx. USD 4–5 million). The appropriation for this project will be submitted for approval to the *Folketing* Finance Committee. See also note 117.

Concerning the EC Extensification Schemes ("set-aside"), including in particular subsidies for environmentally sensitive areas, cf. First Sub-report of the Structure and Planning Committee (Report 1987/1122). Environmentally sensitive areas are there (p. 67ff.) designated as including salt meadows, fresh-water meadows and areas close to watercourses and lakes, i.e. areas which are also relevant to the Ramsar Convention. Furthermore, in a letter of March 30, 1989 from the Ministry of Agriculture, it is pointed out that development of environmentally sensitive areas (ESA's) should include EC bird protection areas (and Ramsar sites) which are particularly sensitive to intensive agriculture. See Act no. 382 of June 7, 1989. See also A. Hopkins, etc. in Upland Grasslands in England and Wales (Biological Conservation 47, 1989, p. 234).

As an element of follow-up of the EC Extensification Schemes, which i.a. include subsidies for the private afforestation (Council Order (EEC No. 797/85)), on November 17, 1988 a Bill was put forward (L 90) by the Minister for the Environment for amendments to regional planning legislation. According to these amendments regional plans (see note 71 b) must designate afforestation areas, as well as areas where afforestation is not desired. According to the explanatory notes to the Bill (adopted by the Folketing on May 24, 1989) Ramsar sites and EC Bird Protection Areas are areas where afforestation is not desired (with the natural exception of those forest areas designated as EC Bird Protection Areas). For this reason financial subsidies cannot be given for afforestation in these areas. This corresponds to the recommendations of the Second Sub-Report of the Structure and Planning Committee set up by the Ministry of Agriculture (cf. Report 1988/1145, pp. 78 and 90 f.).

69) The definition in the 1984 Circular related to the Executive Order from 1981 was occasioned by, amongst other things, the fact that the same small bathing etties and landing stages had been exempted from the provisions of the Conservation of Nature Act relating to beach protection lines (Section 46, Subsection 1) n the Ministry of the Environment's Executive Order No. 549 of November 16, 1983. For this reason there was no obvious guarantee that conservation would be aken into consideration. The Circular can also be taken to mean that the Ministry of Public Works, now the Ministry of Transport, has felt a continual obligation to ive up to the ratification assumptions. As to breakwaters and other coastal protecion systems (as well as certain cables and pipelines) see Circular No. 168 of October 21, 1980 from the Ministry of Public Works to the Coast Inspection, according to which the National Forest and Nature Agency shall be consulted "if an installation, by virtue of its structure, dimensions or *location* (author's emphasis) ... would imply significant reservations from the aspect of nature conservation." See also the letter of December 22, 1981 from the National Agency for Protection of Nature, Monuments and Sites to the Ministry of Public Works upholding the requirement to remove a fascine which had been established without a permit and was found to conflict in principle with the intentions of the Ramsar Convention.

For the new Coastal Protection Act (Act No. 108 of March 5, 1988 concerning coastal protection) and in particular the consideration of the Ramsar sites (and EC Bird Protection Areas), see the Ministry of the Environment's letter of May 24, 1988 and the letter of September 20, 1988 from the National Forest and Nature Agency to the regional authorities stating that coastal protection works involving significant ecological changes in Ramsar sites (or EC bird protection areas) would be in conflict with international obligations and that cases of doubt should be referred to the National Forest and Nature Agency.

(70) For example letter of April 13, 1983 concerning Vejle, letter of October 17, 1983 concerning Funen and letter of March 10, 1988 concerning Southern Jutland (see note 74).

(71) For this reason amongst others the Department of the Ministry of the Environment in its letter of January 8, 1986 upheld the National Agency of Environmental Protection's refusal of a permit for a salt-water fish farm in Bøgestrømmen, see the Minister for the Environment's memorandum of January 15, 1985 to the *Folketing* Environment and Planning Committee (question no. 130 — genera section 147) on the establishment of aquacultures in Ramsar sites. According to the memorandum, the designation of Ramsar sites implies "that nature conservation considerations should be accorded special weight when balancing the various interests" in deciding upon aquaculture applications. The addition that "the positive commercial significance such installations may have for small island communities" may be among the other interests to be assessed must presumably be taken as an expression of the national interest in preserving these communities.

(71a) The Executive Order concerning lead shot is one of the few available examples of provisions concerning Ramsar sites which do not also apply to the EC Bird Protection Areas which happen not to be Ramsar sites. See note 26 in this connection. For provisions on fur farms in relation to Ramsar sites, see the letter of March 17, 1988 from the National Forest and Nature Agency.

(71b) Generally speaking, one of the basic principles of Danish legislation concerning land use and environmental protection is that any major change in current land use either requires a special licence or must be in accordance with the existing lawful physical or environmental planning. On the other hand, developments incidental to existing land use, e.g. agricultural development, can normally take place without prior authorization.

Furthermore, it is a common trend of the legislation covering different aspects of land use, including legislation on comprehensive physical planning or on sectoral plans (such as on the quality of receiving media, water supply, waste management and raw materials), that environmental impacts must be taken into consideration in the decision-making process, either directly (because the very purpose of the legislation is to protect the environment, including habitats) or indirectly (by the establishment of procedures whereby nature conservation authorities must approve the project/plan, or at least be consulted).

A third important feature of this legislation is the possibility of either public participation in the decision-making process (elaboration of physical plans) or the right of local or national NGOs to appeal to central authorities against local and regional decisions.

Planning Legislation:

National and Regional Planning Act (Promulgation Order No. 735 21/12 1982 with later amendments) providing for a comprehensive physical planning system both on the national scale and at the regional level (Denmark is divided into 14 counties, often referred to as regions).

Greater Copenhagen Planning Act (Promulgation Order No. 736 21/12 1982 with later amendments) providing for a comprehensive physical planning system in the Metropolitan area (the 3 counties of Copenhagen, Frederiksberg and Roskilde).

Municipal Planning Act (Promulgation Order No. 391 22/7 1985 as amended). This Act provides for two types of plan: A municipal structure plan (Denmark is divided into approx. 250 municipalities) and local plans.

The municipal structure plan, containing the framework of local planning, must not be contrary to regional planning or national planning directives, such as e.g. the prohibition on building holiday cottages and hotels in the coastal zone as defined in Circular Letter No. 167 (28/8 1981) from the Ministry of the Environment. A local plan, which also ultimately depends on the regional plan, is needed before major building and construction works, amongstother things, can be undertaken.

A special area may be protected as a habitat under a local plan, but this is normally appropriate only for small areas of local interest and situated inside — or in the vicinity of — towns. In such cases, the municipality is the obvious body to be responsible for the maintenance of the area.

Urban and Rural Zones Act (Promulgation Order No. 446 3/10 1985 as amended) dividing the country into three types of zone:

— urban zones
— holiday cottage districts and
— rural zones.

Rural zones are a residual category, which can only be transferred to another category by a local plan. In rural zones, the construction of new buidings, etc., is generally not allowed without special permission.

Sectoral plans are of a mandatory character insofar as they are incorporated in regional plans and are required e.g. by *the Environmental Protection Act* (Promulgation Order No. 85 8/3 1985 with later amendments), the *Conservation of Nature Act* (Promulgation Order No. 530 10/10 1984 as amended by Act No. 355 13/5 1987), and the *Raw Materials Act* (Promulgation Order No. 617 24/9 1987).

The legislation mentioned above is described in detail in Ole Christiansen: Comprehensive Physical Planning in Denmark (pp. 70–105 in: Planning Law in Western Europe, Elsevier, North-Holland, 1986). Reference is also made to Klaus Illum in: European Environmental Yearbook, Docter, Milan 1987, pp. 538–540.

(72) Cf. Ole Christiansen and Inger Vaaben in Landsplanlægning ("Nationa Planning") (Juristen 1983) p. 135 and p. 149, note 6. For regional planning see also the first Sub-report of the Land Committee (1985/1051) p. 15ff.

(73) The Ministry of the Environment's letter of January 7, 1987. For the Minister's approval of regional plans and additions thereto ("passive national planning"), see Ole Christiansen and Inger Vaaben in Juristen 1983, p. 143f. For Ramsar sites in relation to large wind turbines and wind farms, see Large Wind Turbines in Denmark (English summary) (National Agency for Physical Planning 1985) pp. 21ff. and 44, according to which wetlands falling under the Ramsar Convention (and EC Bird Protection Areas) should in principle be kept free of wind farms. Such installations have furthermore to some extent been the subject of special regional planning initiatives, see e.g. the Regional Plan for North Jutland Addendum No. 8 (January 1987) where the precondition for certain designations is that Ramsar sites amongst others are kept exempt (p. 10). See also letter of July 1, 1988 concerning EC Bird Protection Areas. Regional Planning Addendum No. 12 from the County of Funen concerning 3 wind farms could not be recommended by the National Forest and Nature Agency (letter of January 8, 1987) because of its impact on Ramsar sites. In the Ministry of the Environment's approval of the Regional Planning Addendum (letter of August 9, 1988), two of the sites were excluded on this basis, while for another site "crucial importance was attached to the safeguarding of ornithological aspects, including the fact that bird migration from surrounding water bodies designated as EC Bird Protection Areas is not hampered". In a letter of March 6, 1984 from the National Agency for Protection of Nature, Monuments and Sites, an application for a wind farm was refused because its potential location bordered immediately on a Ramsar site (see note 8). The National Forest and Nature Agency's standpoint was upheld by the Minister for the Environment. From the Department's letter of April 17, 1986 it does, however, appear that this does not express a general prohibition on the establishment of wind turbines in these areas, since it might be possible to erect small wind turbines. The guidelines (in Chapter 6.3.8.) in Circular No. 87 of June 29, 1984 for the granting of permits for small wind turbines thus also applies for the Ramsar sites, at any rate. Regardless of the above circular, there are several regional planning guidelines, under which small wind turbines may generally not be erected in special protection and similar areas. This is in accordance with the view of the National Forest and Nature Agency, that small wind turbines (singly or in pairs) should only be permitted in Ramsar sites if this is acceptable on a case-by-case evaluation, and large wind turbines or wind farms should generally be prohibited in Ramsar sites, see the Agency's letter of January 22, 1987.

In a letter of January 26, 1989 that National Forest and Nature Agency did however, accept a large wind turbine in a Ramsar site on the island off Læsø, due to difficulties in finding alternatives and taking into consideration that the wind turbine was located in an agricultural area outside the most valuable habitats in the Ramsar site.

In the Agency's view, there is a reverse burden of proof, so that the applicant must establish that the installation will not be detrimental to the site. See in this respect Ellen Margrethe Basse: Miljøankenævnet ("The Environmental Appeal Board") (Gad, 1986), p. 376, according to which "the Appeal Board will virtually always grant permission for the establishment of small wind turbines". For practice for wind turbines, see also Chapter 5.6.4.

(74) The Regional Planning Document has been drawn up partly on the basis of a memorandum from the National Agency for Protection of Nature, Monuments and Sites of September 24, 1986 concerning Danish administration of the Convention so far.

The Ministry of the Environment's approval of Regional Planning Supplement No. 3 included as an Annex to the plan, contains diffrent conditions, i.a. concerning Ramsar sites and EC Bird Protection Areas, e.g. concerning wind turbines in the Wadden Sea and dinghy berths. See Chapter 5.4.2. and note 70. Extension of dinghy berths in the Wadden Sea may not take place.

(75) In the explanatory notes, reference is also made to the fact that the designation includes land areas proposed by the *Folketing* Environment and Planning Committee. This refers to several memoranda to the Committee from the Minister for the Environment concerning notification of the Wadden Sea as a Ramsar site, most recently of March 13, 1984 (re questions nos. 3 and 4 — general section, Annex 13) where the delineation of the potential new Ramsar site was outlined and the proposed procedure was explained, comprising more detailed delineation in a regional planning context. The effects of notification as a Ramsar site described in the County of Ribe's Regional Plan is based to a certain extent on this memorandum. See also the parliamentary question in the Official Report of the *Folketing* Proceedings 1984–85, column 7629ff. (question no. S 650).

(76) This, to a certain extent also concerns the Environmental Appeal Board, see Bent Christensen et al. in Rammestyring i Miljøretten ("Framework Control in Environmental Law") (Juristen 1984) p. 185, in accordance with special provisions in various acts under the jurisdiction of the Ministry of the Environment. See, in some respects, to the contrary note 101. See also Ole Christiansen and Inger Vaaben, op. cit. (note 72) p. 144.

(77) Miljørettens Grundbog ("A Textbook of Environmental Law"), op. cit. (note 1), p. 242. See also Miljøordbog (environment lexicon) (the Association of County Councils in Denmark, 1986) where Ramsar sites are laconically defined as "wetlands of international importance for waterfowl. *The sites are determined by an international convention* (the Ramsar Convention)" (author's emphasis).

(78) For fluid planning, cf. Bent Christensen, op. cit. (note 76) p. 177f. and the remarks in the Official Report of the *Folketing* Proceedings 1986–87, Addendum A, column 343 ff to Section 5, no. lb in Act No. 355 of May 13, 1987 concerning the amendment of various planning laws.

(79) Ole Christiansen in Nordisk Administrativt Tidsskrift 1985, p. 316f. (also published as Rammestyringssystemet ("The Framework Control System" (special edition no. 1, National Agency for Physical Planning, 1986)).

(80) Ole Christiansen, op. cit. (note 79), p. 316f.

(81) As an appeal authority, there are no constraints on the Agency, cf. Bent Christensen et al. op. cit. (note 76) p. 184. However, the Agency has been involved in the process of approval of the plan and will at times "feel bound", not only where only a right of protest (local plans) is involved, but also where actual powers exist (however, cf. otherwise in the example mentioned in note 100), cf. in part Ole Christiansen, op. cit. (note 79) p. 318 according to which "it can be claimed that to some extent the plan forces those exercising discretionary powers to respect plans".

(82) Cf. Karsten Revsbech: Planer og Forvaltningsret ("Plans and Constitutional Law") (Gad, 1986)

(82a) See Niels Madsen, op. cit. (note 17), p. 39, according to which treaty law obligations can involve disregarding provisions in circulars.

(82b) See on this question Bendt Andersen and Ole Christiansen: Kommuneplanloven ("Municipal Planning Act") (Juristforbundets Forlag, 1989), p. 302ff. and as to protests on legal questions in cases where no period for objections is provided for, p. 528ff. on Section 48 of the Municipal Planning Act.

Under Section 26, Municipal Planning Act, if a government body enters a timely objection to a local planning proposal relating to special interests for which it is responsible, the Municipality may not finally adopt the plan until agreement has been reached between the Municipality and the government body on the content of the plan. If agreement is not reached the question may be referred to the Minister for the Environment.

(83) The guidelines in Fredningsplanlægning ("Conservation Planning") No. 1 (National Agency for Protection of Nature, Monuments and Sites, 1980) p. 32, see Circular No. 26 of February 7, 1980, items 4 and 6, and Guidelines in Fredningsplanlægning ("Conservation Planning") No. 2 (National Agency for Protection of Nature, Monuments and Sites, 1982) p. 52, cf. Circular No. 171 of October 21, 1982. See also Report 1985/1051 (note 72) p. 79f. Compare also the Regional Planning Act Chapter 4, Subsection 1 and Subsection 2, added by Act No. 355 of May 13, 1987.

34) Guidelines for Recipient Quality Planning, Part II/Coastal Areas (in Danish) National Agency of Environmental Protection, 1983) pp. 19f. and 99. See e.g. Recipient Quality Plan for Vejle County (in Danish) (Vejle County Council 1985) p. 4 (here it is also assumed "that new polluting activities will not be established in" he Ramsar site), Recipient Quality Plan for the County of Storstrøm (in Danish) 1985), e.g. Part 3: Lakes, p. 18 and Part 4: Coastal Waters, e.g. pp. 25, 30, 34, 9, 51 and 55, and Recipient Quality Plan 1 for the County of West Zealand (in Danish), e.g. pp. 68, 98, 100 and 102. In the Recipient Quality Plan for the Lim iord (in Danish) (the Lim Fiord Committee, 1986), the stricter objective is not pplied at all, however, due to the level of pollution in the area, i.e. not for the Ramsar sites situated there either, cf. pp. 5, 7 and 100.

35) See Ellen Margrethe Basse, op. cit. (note 63), p. 698, note 12.

36) As an example see Havbundsundersøgelser — Råstoffer og fredningsineresser. Sjællands Rev ("Seabed Surveys — Raw Materials and Conservation nterests. The Zealand Reef") (the National Forest and Nature Agency, 1987) pp. –10 and 57. Extraction of raw materials on the seabed and fisheries are the least egulated activities in relation to the Danish Ramsar sites. The extraction of raw naterials on the seabed is in principle without restriction for the vessels holding xtraction permits. This, in principle, means unrestricted extraction in Ramsar ites, subject to certain general limitations (see the aforementioned publication, p. 9–10). The philosophy hitherto has been that the extraction of raw materials enerally does not harm biological interests and that, should this exceptionally be he case, practical measures must be taken. This philosophy is perhaps not ompletely correct. In relation to the Ramsar Convention, the purpose of the Convention is not only to protect birdlife but also wetlands as such, including flora nd fauna on the seabed. Special conservation measures for Ramsar sites in erritorial waters regulate or prohibit the extraction of raw materials. For the lational Forest and Nature Agency's "policy" on exploitation of the seabed, see he folder "Great Riches are Hidden in the Sea" (English edition) (1988).

37) The Minister for the Environment's Report of May 28, 1986 to the Folketing Environmental and Planning Committee (questions nos. 204 and 205 — general ection, Annex 298). See also Pelle Andersen-Harild: Danish Oil Drilling Activities n Relation to the Wadden Sea (Proceedings of the Fifth International Wadden Sea Symposium, the National Forest and Nature Agency, 1987), p. 292f.

38) See generally Ole Christiansen, op. cit. (note 79), pp. 316–318.

n a theoretical example, "Nibe Municipality — Municipal Plan 1988–97", eviewed in the Municipal Plan for Rural Areas (National Agency of Physical Planning, 1987) (in Danish) Ramsar obligations are respected but on the other and the unfortunate trend towards wind farms being situated just outside Ramsar sites is apparent, cf. p. 25 and pp. 28–29. Compare with note 73.

(89) A proportion of the aforementioned cases concern the period prior to the sc
called normal addendum 1984/1985 to the Regional Plans, which extends thes
plans to 1997 and in particular features an amplification of planning for the ope
landscape, cf. e.g. the approximate size and location of yachting harbours (se
Guidelines in Regional Planning No. 5 (National Agency for Physical Planning
1983) (in Danish), p. 21).

(90) The National Agency for Protection of Nature, Monuments and Sites' letter
of August 15, 1984 and April 19, 1985. Cases concerning harbour installation
can also be submitted by means other than local planning cases. See the Chie
Conservation Board's decision of January 8, 1981 concerning the removal of
harbour installation established in a Ramsar site without authorization accordin
to Section 46, Subsection 1 of the Conservation of Nature Act. The Chief Conse
vation Board attached importance inter alia to the circumstance that navigation i
the area would have been "incompatible with the considerations which have led t
the adoption of the Ramsar Convention".

(91) The National Agency for Protection of Nature, Monuments and Sites' lette
of April 1, 1980 in which an objection was made.

(92) The National Agency for Protection of Nature, Monuments and Sites' letter
of November 22, 1978 and of November 23, 1982.

(93) The National Agency for Protection of Nature, Monuments and Sites' cas
no. F 1401–13 and letter of January 25, 1983.

(94) The National Agency for Protection of Nature, Monuments and Sites' lette
of December 17, 1984.

(95) The National Agency for Protection of Nature, Monuments and Sites' lette
of February 20, 1986 and the National Forest and Nature Agency's letter o
September 14, 1987. In cases of this nature, amongst others, there is usuall
close cooperation with the Wildlife Administration of the Ministry of Agriculture
whose views are normally similar to those of the National Forest and Natur
Agency (see note 114). See also note 100. The Wildlife Administration (Huntin
and Game Management Act) has from July 1, 1989 been transferred to th
Ministry of the Environment (National Forest and Nature Agency).

(96) The National Agency for Protection of Nature, Monuments and Sites' lette
of July 23, 1981. See also the National Forest and Nature Agency's letter of Apr
21, 1988 where, on the one hand, a restricted extension of the harbour capacit
for keelboats was approved, while on the other hand the Agency was unable t
recommend an increase in the number of dinghy berths, as these were found t
be contrary to "efforts hitherto to relieve this very sensitive section of the Ramsa
site" (the dinghies were able to sail into an inlet, which was not navigable b
keelboats).

(97) The National Forest and Nature Agency Case No F 245/5–296.

(98) See as an example the case concerning Fjand Hotel at Nissum Fiord as described p. 8ff. in "jeg troede vi havde en aftale. Miljøkonventioner og dansk virkelighed ("I Thought We Had an Agreement. Environmental Conventions and Danish Reality") (Niche, 1985), and as explained in the Minister for the Environment's report of September 2, 1985 to the *Folketing* Environment and Planning Committee (question no. 416/general section — Annex 539).

(99) Cf. the conclusion in Chapter 5.5.3.

(100) Cf. the statements of the Minister for the Environment in the case cited above (and Consultation Memorandum of September 3, 1985, re general section — Annex 559) according to which installations of this nature would not normally be approved in Ramsar sites. In that case, however, the building was approved for various reasons while in the following case (Aale Fælleshegn) approval was refused, cf. the Ministry of the Environment's letter of March 4, 1986. In the case SN 245/5–488, a community holiday project at Starreklint immediately outside the boundary of a Ramsar site was approved. On the other hand, a camping site which had previously been situated within a Ramsar site was simultaneously moved to outside the site. The National Forest and Nature Agency did, however, object to a local planning proposal refusing to accept a projected bathing jetty and bathing islet due to potential disturbance to birdlife in the Ramsar site. See generally the Minister for the Environment's reply of April 18, 1988 to parliamentary question no. S 880. Compare in this respect the Chief Conservation Board's refusal of camping activities seaward of the beach protection line in a Ramsar site, cf. letter of April 14, 1987. In a letter of June 28, 1988 the National Forest and Nature Agency on the other hand accepted the extension of a camping business landward from the beach protection line in a Ramsar site (and EC Bird Protection area), but simultaneously requested the Conservation Board to fence off the outermost 300 m of a promontory during the birds' breeding season. See also Chapter 5.6.2.

(101) The review in Ellen Margrethe Basse: Miljøankenævnet ("The Environmental Appeal Board") (note 63) must be read together with the general conclusions on the subject of the Urban and Rural Zones Act (p. 342ff.), that Country Planning Directives and Regional Plans are given higher priority (by the Appeal Board) but not construed as binding (p. 350 and p. 345) and that the Appeal Board takes into account conservation authorities' evaluations but, despite the fact that particularly weighty landscape considerations are normally given higher priority than others, does not feel bound in this respect, cf. also note 76 above (p. 353 and p. 680, note 50. Compare also p. 690, note 137). One of Ellen Margrethe Basse's general conclusions is also that the Environmental Appeal Board's practice shows that ecological considerations are often given very low priority, op. cit. p. 518.

(102) The Environmental Appeal Board's evaluation of its own practice is not quite so rigorous, cf. the decision of July 31, 1986 that "the Board, in accordance

with the guidelines expressed in ... Urban and Rural Zones Act Circular ... has followed the practice of normally granting permission for the establishment of small wind turbines, unless particularly weighty landscape considerations are decisively against". Compare note 73 above.

(103) Decision of May 17, 1982 is referred to by Ellen Margrethe Basse, op. cit. (note 63) p. 353 (cf. note 50, p. 680) and p. 377 (cf. note 137, p. 690). See also Rulings Concerning Real Property (in Danish) (KFE) 1983, note 1, p. 183.

(103a) In a ruling of February 6, 1989 (5065/7/22–26) the Environmental Appeal Board has permitted two wind turbines (of 145 kW) in Ramsar site no. 17. The reason is that the erection of the wind turbines did not conflict with the landscape resources and that the turbines are located on the edge of the Ramsar site (and EC Bird Protection Area). The National Forest and Nature Agency had pronounced the opinion that the erection of the wind turbines would not be compatible with international obligations. The Environmental Appeal Board's ruling must be considered to be wrong, cf. note 73. Furthermore, it is very unfortunate in legal terms, since in reality the Appeal Board suppresses the existence of the international obligations by which the Board is also bound.

(104) See the Environmental Protection Act, Section 75(2), concerning the Board's independence of instructions. Although the Board is naturally bound by legislation, etc., cf. note 76.

(105) Compare also the Chief Conservation Board's unanimous refusal of an application for the erection of 5 wind turbines in an area affected by the beach protection line in Onsbjerg Parish, Samsø. The area did not lie in a Ramsar site, but the Chief Conservation Board's ruling (letter of September 10, 1985), which in principle attaches great importance to society's interests in alternative sources of energy (cf. Ellen Margrethe Basse, op. cit. note 63, p. 689, note 132), is partially based on a statement from the National Agency for Protection of Nature, Monuments and Sites that the 5 wind turbines would constitute a barrier between two Ramsar sites, with the risk that birds would fly into the turbines.

(106) This ruling resembles the decision of February 6, 1985 quoted in Ellen Margrethe Basse, op. cit. (note 63) p. 478 concerning dumping of unpolluted moraine deposits in a Ramsar site.

(107) On the decision, cf. also Ellen Margrethe Basse, op. cit. (note 63) p. 382. Compare note 149, p. 691. The statement here, that "only extraordinarily important conservation interests can conceivably be given higher priority than public works" is perhaps not incorrect viewed in the light of the general practice of the Environmental Appeal Board. On the other hand, it is misleading when considered in a broader context. And in relation to the Ramsar Convention the starting point is presumably that there must be strong national interests attached to a public installation before it can be given higher priority than the Ramsar obligations.

(107a) Cf. "Information from the Chief Conservation Board" (in Danish) Nos. 412 A. In No. 412 B a decision concerning the planting of hedgerows in a Ramsar site is referred to. The planting was permitted despite the fact that the area was an important resting place for geese, but the cultivation was on a modest scale and the National Forest and Nature Agency had not stated to the Chief Conservation Board that the planting was in conflict with the Convention.

(108) Cf. the comment in note 3 in Rulings concerning Real Property (in Danish) (KFE) 1982 p. 213ff. concerning the Chief Conservation Board's ruling of November 17, 1981 on the conservation of Stavns Fiord, according to which the Convention calls on Denmark to promote conservation.

(108a) Conservation action can be taken either under Chapter III or Section 60 of the Conservation of Nature Act. See also note 67.

Conservation action under Chapter III, which applies only to rural areas, takes place in accordance with aspecial public procedure and is determined by the Conservation Board for the region in which the land is situated (there are approx. 20 conservation boards in total). The decision (if taken) contains provisions on conservation and compensation to land owners, since conservation is considered to be in the nature of expropriation, and so can only take place on payment of full compensation, cf. Ole Christiansen: Comprehensive Physical Planning in Denmark (op. cit. note 1) p. 70 and Veit Koester in Environmental Policy and Law (op. cit. (note 1, Chapter A) p. 110f.). The decision can be submitted to the central Chief Conservation Board, and this is mandatory for major cases.

Under Section 60, Conservation of Nature Act, the Minister for the Environment may protect State-owned land areas and areas in territorial waters by executive order. No special procedure is stipulated in this respect and conservation action takes place without compensation. Under the Bill, L No. 213, mentioned in note 62, it is proposed to extend the power of protection by means of an executive order to fishing zones.

(108b) According to information from the Inspector of Game Reserves, Mr. Palle Uhd Jepsen, the Wildlife Administration.

(108c) As to the Act of 1972 see also note 117. Land areas within Ramsar sites are predominantly in private ownership. 11–12 pct. of the areas are today in State ownership.

In connection with parliamentary debate of the 1988 Budget the Parliamentary Finance Committee from the Minister for the Environment requested a list of state-owned properties which might be sold as an element of the Government's privatization measures. In the Minister's report on this issue to the Finanace Committee of November 21, 1988 the properties which are situated in i.a. Ramsar sites and EC Bird Protection Areas are expressly excluded from the list (cf. p. 49 of Annex 3 to the Supplementary Report on the Budget Act 1989 proposals (in Danish)).

(108d) Digging for cockles requires a permit from the National Forest and Nature Agency under the Wadden Sea Conservation Order. On the other hand, the taking of common mussels in the Wadden Sea is regulated by a special executive order (no. 872 of December 22, 1988) issued by the Danish Ministry of Fisheries. Under this order, the taking of common mussels is considerably restricted in terms of quantities and is completely prohibited in certain places, e.g. Ho Bugt where the Wadden Sea scientific research area is situated.

(109) The Chief Conservation Board's decision of November 23, 1984 concerning Sønderholme and Plet Enge, for which specific conservation action was decided on, despite the fact that the area was subject to general protection under the Conservation of Nature Act, Section 43b (see Chapter 5.3.1.), cases in which the Chief Conservation Board is usually very reticent in carrying out further conservation. See also as to this decision Rulings Concerning Real Property (in Danish) (KFE) 1985, p. 87, note 3, and concerning the general problem Karnov 11th edition p. 1368, 2nd column in the middle (note 205). The Chief Conservation Board's decision in a similar case concerning Horskær on Gylling Næs on May 10, 1985 is in line with the position stated in the text. The most recent conservation order has been accomplished by the Chief Conservation Board's decision of October 25, 1988, concerning approx. 195 hectares in Dybsø Fiord (Ramsar site no. 20).

(110) See Rulings Concerning Real Property (in Danish) (KFE) 1979 p. 117ff concerning the Chief Conservation Board's decision of October 5, 1978 on the conservation of Borreby and Østerhovedgård, cf. Ellen Margrethe Basse, op. cit (note 63) p. 623, note 84 (and 82) and p. 678, note 31. For subsequent follow-up of this conservation case, see the decision of February 9, 1987 concerning Stigsnæs (see the Chief Conservation Board's Current Information Newsletter Information from the Chief Conservation Board, no. 403 (in Danish)) the Ramsar Convention had a "reverse" effect since some areas which were not included in the delineation of the Ramsar site were omitted from conservation, also with reference to the Regional Plan. Further reference can be made to KFE 1984, p 158ff. concerning the Chief Conservation Board's decision of December 27, 1983 on the conservation of Saltholm against the wishes of the Greater Copenhagen Council and the County Council of Copenhagen. This case did not involve a Ramsar site, but an EC Bird Protection Area. See also references in Karnov 11th edition, p. 1359, note 44, and "Information from the Chief Conservation Board" (in Danish) Nos. 330 and 412 c. The Chief Conservation Board by and large takes the same position on country and regional planning as the Environmental Appeal Board (cf. note 101), which, like conservation planning, is not binding on the Conservation Board, cf. Report 1985/1051 (note 72) p. 80.

(111) In the case described in note 108 certain limitations were stipulated on the general public's access, with reference to the fact that its exercise might otherwise "be considered to be incompatible with the considerations which have led to the adoption of the Ramsar Convention". In a ruling from 1989 (the Chief Conservation Board's ruling of May 19, 1989) a significant proportion of the land areas in

amsar Site No. 25 (EC Bird Protection Area No. 83/Hyllested-Rødsand) was
nade subject to a conservation order "with reference to Denmark's international
ommitments and the national planning pursuant thereto". (Compare with Chap-
er 5.2.2. concerning the regional plan for the County of Storstrøm). Due to the
irdlife, the conservation rules contain a ban on new hedgerows and the renewal
f existing hedgerows. In uncultivated areas, meadows and dry grassland, fertiliz-
rs, herbicides and pesticides may not be used.

112) The Chief Conservation Board's decision of November 12,1985 concerning
ne protection of Margrethe-Kog, cf. Rulings Concerning Real Property (in Dan-
sh) (KFE) 1986, p. 85ff. Also in this case a large area under State ownership was
lso protected, despite the fact that it was not referred to in the County Council
nd the National Agency for Protection of Nature, Monuments and Sites' claim as
ne appellant in the case, cf. KFE 1986, p. 85, note 8. Viewed in relation to this the
tatement (from the Chairman of the Chief Conservation Board) in KFE 1984 p.
59, note 2, that "the Chief Conservation Board ... in various conservation cases
as attached considerable emphasis to the relevant area falling under the Ram-
ar Convention", is at all events not exaggerated.

113) Cf. Section 4, Conservation of Nature Act, according to which there is a
hairman who is a lawyer, 2 members appointed by the Supreme Court from
mong the members of the Court and 1 member from (and elected by) each of the
arties represented in the Folketing Finance Committee.

114) The fact that the role played by the Ministry of Agriculture's Wildlife
dministration in the safeguarding and protection of the Ramsar sites is only
inted at (Chapter 4.2. and notes 32 and 95) is due to the fact that the compe-
ence of the Wildlife Administration, with the exception of game reserves (cf.
hapter 6.1.) is quite limited. In practice the Wildlife Administration plays a very
xtensive role, including acting as authority of first instance in regional as well as
entral administrative contexts. In by far the majority of cases, the views of the
Vildlife Administration and the National Forest and Nature Agency concur. See
ote 95.

114a) A 1988/1989 analysis of two Ramsar sites' development (Lars Rudfeld
nd Morten Rasmussen: Ramsar-områders udvikling og administration siden
978 — en empirisk analyse (Ramsar Sites' Development and Administration
ince 1978 — An Empirical Analysis). Skov-og Naturstyrelsen, 1989 shows the
ollowing:

n Ramsar Site no. 2 (Ringkøbing Fiord), water quality has deteriorated signific-
ntly and it has been possible to prove connections between this deterioration
nd fluctuations in bird stock. On the other hand, data for Ramsar Site no. 22
Nyord/Ulfshale, Præstø Fiord and Fed, etc.) was insufficient to prove links be-
veen development in bird stock and pollution.

Use of land has been analysed on the basis of aerial photographs and inspectio
of the archives of municipalities and county boards, etc., from which it emerge
that significant changes have taken place but that these have usually been insig
nificant to birdlife. It is furthermore concluded that no major landscape change
have taken place (cultivation, overgrowth, planting, etc.).

In the latter respect the result of the empirical analysis by and large correspond
to the conclusions in Chapter 7 of this legal analysis, which are that the areas
status as Ramsar sites has by and large involved a kind of status quo protection i
relation to use and exploitation of land.

On the other hand, the deterioration in Danish Ramsar sites which has take
place on the basis of deterioration in the pollution status of marine areas has no
been counterbalanced by their status as international areas worthy of conserva
tion (however, see Chapter 5.5.5. on the more stringent objectives of recipien
quality plans in Ramsar sites). However, this is not surprising. In a country a
small as Denmark, with many, and relatively large, Ramsar sites, the pollutio
status of Ramsar sites is naturally dependent on the general pollution status o
marine areas. Improved water quality in Danish Ramsar sites therefore cannot –
generally speaking — be achieved without general improvement in all Danis
marine areas.

(115) See Chapter 4.5.3. and Niels Madsen, op. cit. (note 17) p. 40 and Ol
Espersen, op. cit. (note 42) p. 375ff. Cf. also note 60.

In view of the intervening development in respect of the accomplishment of rule
of law, the development of practice in individual cases and the current plannin
stage, the correct time for a general National Planning Directive concerning th
Ramsar Convention (and the EC Bird Protection Areas) must be considered t
have passed, at any rate for the time being. All in all, I would not hesitate to stat
that in the present circumstances such a Directive would do more harm than goo
and would at best be without any impact whatsoever.

(116) The Convention can, as stated, influence Danish legal application (cf. not
50) and it must therefore be possible for private individuals to invoke it vis-à-vi
the courts in a case where there is an independent right of legal action, cf. th
conclusions of Niels Madsen, op. cit. (note 17) p. 39, and Claus Gulmann, op. ci
(note 46) p. 49f., and in "Juristen" 1988 p. 287f.

(117) In some parts of the text (cf. e.g. Chapter 5.3.5.) and of the notes (e.g
notes 18, 25 and 62) reference is made to various Bills which lapsed when th
Parliamentary Elections were called in April 1988.

Most of these Bills have been re-submitted to the *Folketing* in the autumn of 198
(See note 68 on some Acts passed by the *Folketing* in June, 1989). However, thi
does not include the Conservation of Nature Act, since the proposed amend
ments will be included in a revised Conservation of Nature Act which will b

resented to the *Folketing* as a Bill in the autumn of 1989 in connection with a eneral revision of the environmental and planning legislation.

n May/June 1989 two other important Acts were adopted, i.e. a new Forest Act Act No. 383 of June 7, 1989) and a Nature Management Act (Act No. 339 of May 4, 1989), which relates to the overall strategy described in note 68 in respect of narginal land.

he Forest Act replaces the old Act of 1935. The main objective of the Act is to onserve the Danish forests while at the same time upgrading production, wildlife nd enviromental considerations in the forests. The Forest Act also protects the mall biotopes in the forests, e.g. bogs below 5,000 sq.m. and therefore not rotected by Section 43 of the Conservation of Nature Act (see note 61).

he Nature Management Act replaces Act No. 230 of 1972 described in Chapter .1. on the acquisition of real property for recreational purposes. The new Act rovides for care and restoration of nature, i.a. through agreements with private wners, as well as for afforestation and support for outdoor life. The Goverment's nvironmental investment plan allocates DKK 680 million (approx. USD 95 milion) for the fulfilment of the objective of the Act for the period 1989–1992. The roject mentioned in note 68 concerning the restoration of the Skjern Å, Denmark's largest river, is expected to be implemented by means of financial esources allocated under the new Act. Furthermore, implementation of several ther projects relating to wetlands, of which many are situated within the Ramsar rea, is contemplated.

Fishing hamlet at Hejlsminde Nor in the Lillebælt (Ramsar site no. 15) — an nportant passage area for diving ducks.

9. Annexes

9.1 Convention on Wetlands of International Importance especially as Waterfowl Habitat

Ramsar, 2.2.1971
as amended by the Protocol of 3.12.1982

The Contracting Parties,

Recognizing the interdependence of Man and his environment;

Considering the fundamental ecological functions of wetlands as regulators of water regimes and as habitats supporting a characteristic flora and fauna, especially waterfowl;

Being convinced that wetlands constitute a resource of great economic, cultural, scientific, and recreational value, the loss of which would be irreparable;

Desiring to stem the progressive encroachment on and loss of wetlands now and in the future;

Recognizing that waterfowl in their seasonal migrations may transcend frontiers and so should be regarded as an international resource;

Being confident that the conservation of wetlands and their flora and fauna can be ensured by combining far-sighted national policies with co-ordinated international action;

Have agreed as follows:

Article 1

1. For the purpose of this Convention wetlands are areas of marsh, fen, peatland or water, whether natural or artificial, permanent or temporary, with water that is static or flowing, fresh, brackish or salt, including areas of marine water the depth of which at low tide does not exceed six metres.
2. For the purpose of this Convention waterfowl are birds ecologically dependent on wetlands.

Article 2

1. Each Contracting Party shall designate suitable wetlands within its territory for inclusion in a List of Wetlands of International Importance, hereinafter referred to as "the List" which is maintained by the bureau established under Article 8. The boundaries of each wetland shall be precisely described and also delimited on a map and they may incorporate riparian and coastal zones adjacent to the wetlands, and islands or bodies of marine water deeper than six metres at low tide lying within the wetlands, especially where these have importance as waterfowl habitat.
2. Wetlands should be selected for the List on account of their international significance in terms of ecology, botany, zoology, limnology or hydrology. In the first instance wetlands of international importance to waterfowl at any season should be included.
3. The inclusion of a wetland in the List does not prejudice the exclusive sovereign rights of the Contracting Party in whose territory the wetland is situated.
4. Each Contracting Party shall designate at least one wetland to be included in the List when signing this Convention or when depositing its instrument of ratification or accession, as provided in Article 9.

5. Any Contracting Party shall have the right to add to the List further wetlands situated within its territory, to extend the boundaries of those wetlands already included by it in the List, or, because of its urgent national interests, to delete or restrict the boundaries of wetlands already included by it in the List and shall, at the earliest possible time, inform the organization or government responsible for the continuing bureau duties specified in Article 8 of any such changes.
6. Each Contracting Party shall consider its international responsibilities for the conservation, management and wise use of migratory stocks of waterfowl, both when designating entries for the List and when exercising its right to change entries in the List relating to wetlands within its territory.

Article 3

1. The Contracting Parties shall formulate and implement their planning so as to promote the conservation of the wetlands included in the List, and as far as possible the wise use of wetlands in their territory.
2. Each Contracting Party shall arrange to be informed at the earliest possible time if the ecological character of any wetland in its territory and included in the List has changed, is changing or is likely to change as the result of technological developments, pollution or other human interference. Information on such changes shall be passed without delay to the organization or government responsible for the continuing bureau duties specified in Article 8.

Article 4

1. Each Contracting Party shall promote the conservation of wetlands and waterfowl by establishing nature reserves on wetlands, whether they are included in the List or not, and provide adequately for their wardening.
2. Where a Contracting Party in its urgent national interest, deletes or restricts the boundaries of a wetland included in the List, it should as far as possible compensate for any loss of wetland resources, and in particular it should create additional nature reserves for waterfowl and for the protection, either in the same area or elsewhere, of an adequate portion of the original habitat.
3. The Contracting Parties shall encourage research and the exchange of data and publications regarding wetlands and their flora and fauna.
4. The Contracting Parties shall endeavour through management to increase waterfowl populations on appropriate wetlands.
5. The Contracting Parties shall promote the training of personnel competent in the fields of wetland research, management and wardening.

Article 5

The Contracting Parties shall consult with each other about implementing obligations arising from the Convention especially in the case of a wetland extending over the territories of more than one Contracting Party or where a water system is shared by Contracting Parties. They shall at the same time endeavour to co-ordinate and support present and future policies and regulations concerning the conservation of wetlands and their flora and fauna.

Article 6[1]

1. The Contracting Parties shall, as the necessity arises, convene Conferences on the Conservation of Wetlands and Waterfowl.
2. These Conferences shall have an advisory character and shall be competent, inter alia:
   (a) to discuss the implementation of this Convention;
   (b) to discuss additions to and changes in the List;
   (c) to consider information regarding changes in the ecological character of wetlands included in the List provided in accordance with paragraph 2 of Article 3;
   (d) to make general or specific recommendations to the Contracting Parties regarding the conservation, management and wise use of wetlands and their flora and fauna;
   (e) to request relevant international bodies to prepare reports and statistics on matters which are essentially international in character affecting wetlands;
3. The Contracting Parties shall ensure that those responsible at all levels for wetlands management shall be informed of, and take into consideration, recommendations of such Conferences concerning the conservation, management and wise use of wetlands and their flora and fauna.

Article 7[1]

1. The representatives of the Contracting Parties at such Conferences should include persons who are experts on wetlands or waterfowl by reason of knowledge and experience gained in scientific, administrative or other appropriate capacities.
2. Each of the Contracting Parties represented at a Conference shall have one vote, recommendations being adopted by a simple majority of the votes cast, provided that not less than half the Contracting Parties cast votes.

Article 8

1. The International Union for Conservation of Nature and Natural Resources shall perform the continuing bureau duties under this Convention until such time as another organization or government is appointed by a majority of two-thirds of all Contracting Parties.
2. The continuing bureau duties shall be, inter alia:
   (a) to assist in the convening and organizing of Conferences specified in Article 6;
   (b) to maintain the List of Wetlands of International Importance and to be informed by the Contracting Parties of any additions, extensions, deletions or restrictions concerning wetlands included in the List provided in accordance with paragraph 5 of Article 2;
   (c) to be informed by the Contracting Parties of any changes in the ecological character of wetlands included in the List provided in accordance with paragraph 2 of Article 3;
   (d) to forward notification of any alterations to the List, or changes in character of wetlands included therein, to all Contracting Parties and to arrange for these matters to be discussed at the next Conference;
   (e) to make known to the Contracting Party concerned, the recommendations of the Conferences in respect of such alterations to the List or of changes in the character of wetlands included therein.

---

1 These articles have been amended by the Conference of the Parties on 28.5.1987; these amendments are not yet in force (see separate document).

Article 9

1. This Convention shall remain open for signature indefinitely.
2. Any member of the United Nations or of one of the Specialized Agencies or of the International Atomic Energy Agency or Party to the Statute of the International Court of Justice may become a Party to this Convention by:

   (a) signature without reservation as to ratification;

   (b) signature subject to ratification followed by ratification;

   (c) accession.
3. Ratification or accession shall be effected by the deposit of an instrument of ratification or accession with the Director-General of the United Nations Educational, Scientific and Cultural Organization (hereinafter referred to as "the Depositary").

Article 10[1]

1. This Convention shall enter into force four months after seven States have become Parties to this Convention in accordance with paragraph 2 of Article 9.
2. Thereafter this Convention shall enter into force for each Contracting Party four months after the day of its signature without reservation as to ratification, or its deposit of an instrument of ratification or accession.

Article 10 bis

1. This Convention may be amended at a meeting of the Contracting Parties convened for that purpose in accordance with this article.
2. Proposals for amendment may be made by any Contracting Party.
3. The text of any proposed amendment and the reasons for it shall be communicated to the organization or government performing the continuing bureau duties under the Convention (hereinafter referred to as "the Bureau") and shall promptly be communicated by the Bureau to all Contracting Parties. Any comments on the text by the Contracting Parties shall be communicated to the Bureau within three months of the date on which the amendments were communicated to the Contracting Parties by the Bureau. The Bureau shall, immediately after the last day for submission of comments, communicate to the Contracting Parties all comments submitted by that day.
4. A meeting of Contracting Parties to consider an amendment communicated in accordance with paragraph 3 shall be convened by the Bureau upon the written request of one third of the Contracting Parties. The Bureau shall consult the Parties concerning the time and venue of the meeting.
5. Amendments shall be adopted by a two-thirds majority of the Contracting Parties present and voting.
6. An amendment adopted shall enter into force for the Contracting Parties which have accepted it on the first day of the fourth month following the date on which two thirds of the Contracting Parties have deposited an instrument of acceptance with the Depositary. For each Contracting Party which deposits an instrument of acceptance after the date on which two thirds of the Contracting Parties have deposited an instrument of acceptance, the amendment shall enter into force on the first day of the fourth month following the date of the deposit of its instrument of acceptance.

Article 11

1. This Convention shall continue in force for an indefinite period.
2. Any Contracting Party may denounce this Convention after a period of five years fron the date on which it entered into force for that Party by giving written notice thereof to th Depositary. Denunciation shall take effect four months after the day on which notic thereof is received by the Depositary.

Article 12

1. The Depositary shall inform all States that have signed and acceded to this Convention a soon as possible of:
   (a) signatures to the Convention;
   (b) deposits of instruments of ratification of this Convention;
   (c) deposits of instruments of accession to this Convention;
   (d) the date of entry into force of this Convention;
   (e) notifications of denunciation of this Convention.
2. When this Convention has entered into force, the Depositary shall have it registered wit the Secretariat of the United Nations in accordance with Article 102 of the Charter.

IN WITNESS WHEREOF, the undersigned, being duly authorized to that effect, hav signed this Convention.

DONE at Ramsar this 2nd day of February 1971, in a single original in the Englisl French, German and Russian languages, all texts being equally authentic[2] which shall b deposited with the Depositary which shall send true copies thereof to all Contractin Parties.

---

2 Pursuant to the Final Act of the Conference to conclude the Protocol, the Depositary provided th second Conference of the Contracting Parties with official versions of the Convention in the Arabi Chinese and Spanish languages, prepared in consultation with interested Governments and with th assistance of the Bureau.

9.2 Amendments of the Convention adopted 28 May 1987 by The Extraordinary Conference of the Contracting Parties

Article 6

The present text of paragraph 1 shall be replaced by the following wording:

1. *There shall be established a Conference of the Contracting Parties to review and promote the implementation of this Convention. The Bureau referred to in Article 8, paragraph 1, shall convene ordinary meetings of the Conference of the Contracting Parties at intervals of not more than three years, unless the Conference decides otherwise, and extraordinary meetings at the written requests of at least one third of the Contracting Parties. Each ordinary meeting of the Conference of the Contracting Parties shall determine the time and venue of the next ordinary meeting.*

The introductory phrase of paragraph 2 shall read as follows:

2. *The Conference of the Contracting Parties shall be competent:*

An additional item shall be included at the end of paragraph 2, as follows:

(f) to adopt other recommendations, or resolutions, to promote the functioning of this Convention.

A new paragraph 4 is added which would read as follows:

4. *The Conference of the Contracting Parties shall adopt rules of procedure for each of its meetings.*

New paragraphs 5 and 6 are added, which would read as follows:

5. *The Conference of the Contracting Parties shall establish and keep under review the financial regulations of this Convention. At each of its ordinary meetings, it shall adopt the budget for the next financial period by a two-third majority of Contracting Parties present and voting.*

6. *Each Contracting Party shall contribute to the budget according to a scale of contributions adopted by unanimity of the Contracting Parties present and voting at a meeting of the ordinary Conference of the Contracting Parties.*

Article 7

Paragraph 2 is replaced by the following wording:

2. *Each of the Contracting Parties represented at a Conference shall have one vote, recommendations, resolutions and decisions being adopted by a simple majority of the Contracting Parties present and voting, unless otherwise provided for in this Convention.*

9.3 List of Danish Ramsar Sites, 1987.

1. Fiilsø/County of Ringkøbing
2. Ringkøbing Fiord/County of Ringkøbing
3. Stadil and Veststadil Fiords/County of Ringkøbing
4. Nissum Fiord/County of Ringkøbing
5. Harboøre og Agger tanger
6. Vejlerne and Løgstør Bredning/Counties of Viborg and Northern Jutland
7. Ulvedybet and Nibe Bredning/County of Northern Jutland
8. Hirseholmene/County of Northern Jutland
9. Nordre Rønner/County of Northern Jutland
10. Læsø/County of Northern Jutland
11. Parts of Randers and Mariager Fiords, and the sea adjoining them/Counties of Northern Jutland and Aarhus
12. Sea area north of Anholt Island/County of Aarhus
13. Horsens Fiord and Endelave/Counties of Aarhus and Vejle
14. Stavns Fiord and adjacent waters/County of Aarhus
15. Lillebælt/Counties of Vejle, North Slesvig and Funen
16. Nærå coast and Æbelø area/County of Funen
17. South Funen Archipelago/County of Funen
18. Sejerø Bugt/County of Western Zealand
19. Waters off Skælskør Nor and Glænø/County of Western Zealand
20. Karrebæk, Dybsø and Avnø Fiords/Storstrøm County
21. Waters south-east of Fejø and Femø islands/Storstrøm County
22. Præstø Fiord, Jungshoved Nor, Ulfshale and Nyord/Storstrøm County
23. Nakskov Fiord and Inner Fiord/Storstrøm County
24. Maribo Lakes/Storstrøm County
25. Waters between Lolland and Falster, including Rødsand, Guldborgsund and Bøtø Nor/Storstrøm County
26. The islands Erteholmene east of Bornholm
27. The Wadden Sea

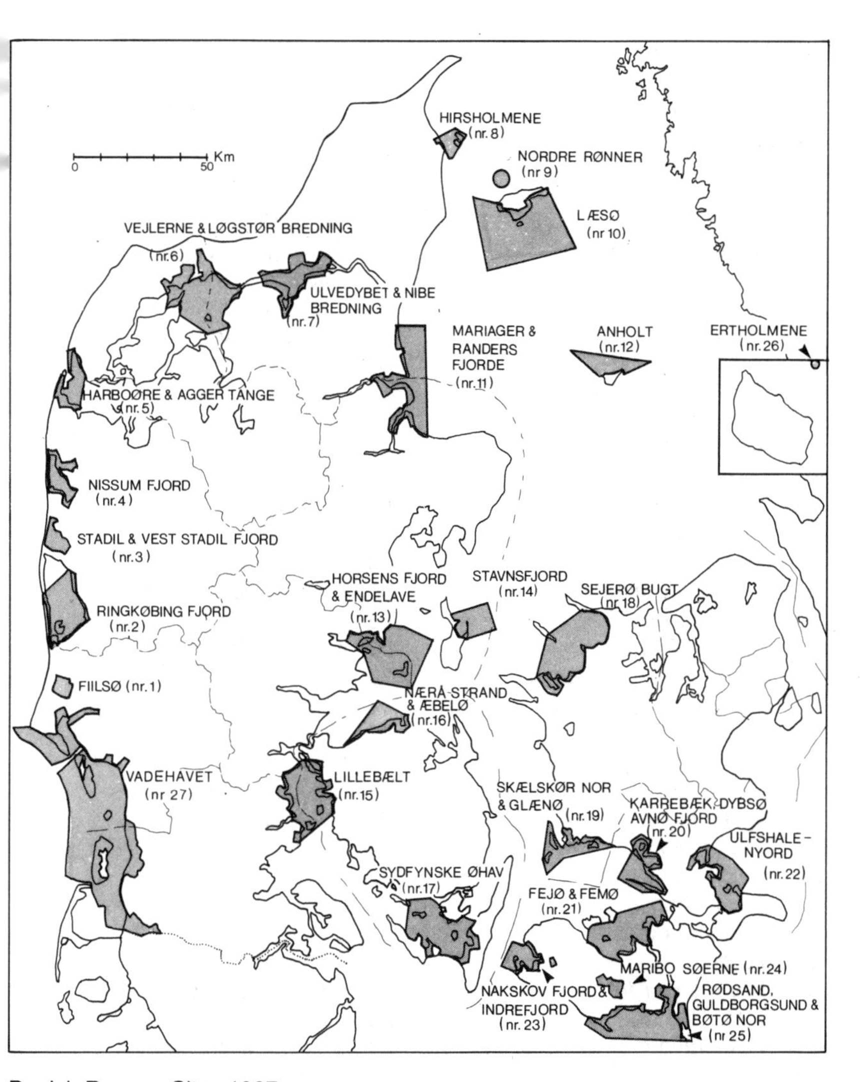

Danish Ramsar Sites 1987

9.4 List of EC Bird Protection Areas in Denmark

1. *Nibe Bredning*
2. Ålborg Bugt
3. Madum Lake
4. Rold Forest
5. Råbjerg Mile
6. Råbjerg Mose
7. Lille Vildmose
8. *Aggersborg*
9. *Nordre Rønner*
10. *Læsø*
11. *Hirsholmene*
12. *Løgstør Bredning*
13. *Østlige Vejler*
14. Lovns Bredning
15. *Randers and Mariager Fiords*
16. Tjele Langsø
17. Ålvand Klithede
18. Vangså Heath
19. *Lønnerup Fiord*
20. *Vestlige Vejler*
21. Ovesø
22. Hanstholm Reserve
23. *Agger Tange*
24. Hjarbæk Fiord
25. Mågerodde, Karby Odde
26. Dråby Vig
27. Agerø
28. Nissum Bredning
29. Flyndersø
30. Kysing Fiord
31. *Stavns Fiord*
32. *Anholt*
33. Salten Langsø
34. Silkeborg Forests
35. Mossø
36. *Horsens Fiord*
37. Borris Heath
38. *Nissum Fiord*
39. *Harboøre Tange*
40. Venø Sound
41. *Stadil Fiord*
42. Sønder Feldborg
43. *Ringkøbing Fiord*
44. Uldum Kær
45. Vejle Fiord Forests
46. Randbøl Heath
47. *Lillebælt*
48. Store Råbjerg
49. *Ho Bugt*
50. Kallesmærsk Heath
51. Ribe Holme, Kongeåen
52. *Mandø*
53. *Fanø*
54. Vejen Mose
55. *Skallingen, Langli*
56. *Fiilsø*
57. *The Wadden Sea*
58. Hostrup Lake
59. Pamhule Forest
60. *Tøndermarsken*
61. Kongens Mose
62. Tinglev Mose
63. Sønder Å-dal
64. Flensborg Fiord
65. *Rømø*
66. Arrild Forests
67. *Ballum Enge*
68. Gråsten Forests
69. Kogsbøl-Skast Mose
70. Frøslev Mose
71. *South Funen Archipelago*
72. Marstal Bugt
73. Vresen
74. Brahetrolleborg Forests
75. Odense Fiord
76. *Æbelø*
77. Romsø
78. Brahetrolleborg Lakes
79. *Erteholmene*
80. Almindingen
81. *Karrebæksminde*
82. *Bøtø Nor*
83. *Hyllested-Rødsand*
84. Grønsund-Ulvsund
85. *Smålandshavet*
86. Guldborgsund
87. *Maribo Lakes*
88. *Nakskov Fiord*
89. *Jungshoved*
90. Klinteskov

91. Holmegårds Mose
92. Vemmetofte Forests
93. Tystrup-Bavelse
94. *Sejerø Bugt*
95. *Skælskør*
96. *Omø-Glænø*
97. Hov Vig
98. Halsskov Reef
99. *Saltbæk Vig*
100. Tissø
101. Bregentved-Gisselfeld
102. Korshage-Hundested
103. Gammel Havdrup
104. Ramsø Mose
105. Roskilde Fiord
106. Arresø
107. Jægerspris Nordskov
108. Gribskov
109. Furesøen
110. Saltholm
111. Vestamager

NB. The areas italicized on the list correspond to Ramsar sites in part or in whole, cf. the Table on the following pages.

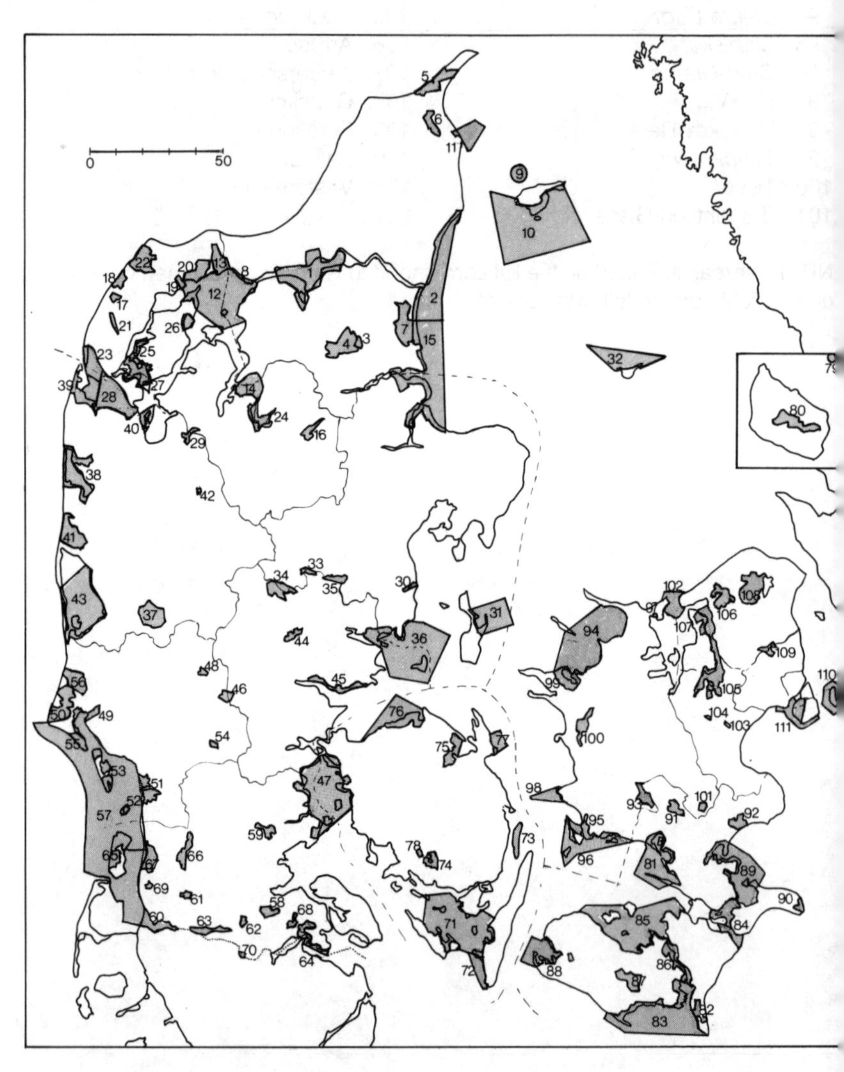

EC Bird Protection Areas in Denmark

9.5 List of Danish Ramsar Sites' Coincidence with EC Bird Protection Areas.

Ramsar Sites	Corresponding numbers of EC areas which are completely included in a Ramsar site	Corresponding numbers of EC areas which extend into the Ramsar Site
1. Fiilsø/County of Ringkøbing		56
2. Ringkøbing Fiord/County of Ringkøbing	43	
3. Stadil and Veststadil Fiords/County of Ringkøbing	41	
4. Nissum Fiord/County of Ringkøbing	38	
5. Harboøre og Agger tanger	23	
6. Vejlerne and Løgstør Bredning/Counties of Viborg and Northern Jutland	8, 12, 13, 19 and 20	
7. Ulvedybet and Nibe Bredning/County of Northern Jutland	1	
8. Hirseholmene/County of Northern Jutland		11
9. Nordre Rønner/County of Northern Jutland	9	
10. Læsø/County of Northern Jutland	10	
11. Parts of Randers and Mariager Fiords, and the sea adjoining them/Counties of Northern Jutland and Aarhus	15	
12. Sea area north of Anholt Island/County of Aarhus	32	
13. Horsens Fiord and Endelave/Counties of Aarhus and Vejle	36	
14. Stavns Fiord and adjacent waters/County of Aarhus	31	
15. Lillebælt/Counties of Vejle, North Slesvig and Funen	47	
16. Nærå coast and Æbelø area/County of Funen	76	
17. South Funen Archipelago/County of Funen	71	
18. Sejerø Bugt/County of Western Zealand	94 and 99	
19. Waters off Skælskør Nor and Glænø/County of Western Zealand	96	95
20. Karrebæk, Dybsø and Avnø Fiords/Storstrøm County	81	
21. Waters south-east of Fejø and Femø islands/Storstrøm County		85

22.	Præstø Fiord, Jungshoved Nor, Ulfshale and Nyord/Storstrøm County	89	
23.	Nakskov Fiord and Inner Fiord/Storstrøm County	88	
24.	Maribo Lakes/Storstrøm County	87	
25.	Waters between Lolland and Falster, including Rødsand,Guldborgsund and Bøtø Nor/Storstrøm County	83	82
26.	The islands Erteholmene east of Bornholm	79	
27.	*The Wadden Sea		57 and 67

* EC areas do not completely cover the Ramsar site.

A colony of Cormorants on Vorsø (Ramsar site no. 13). The Southern Cormorant Phalacrocorax carbo sinensis has increased in number in recent years as a consequence of protection from hunting. In 1987 there were approx. 12,200 pairs in 15 colonies in Denmark, in almost all cases situated in or close to Ramsar sites.

9.6 List of Contracting Parties as of December 1, 1989

S: Signature without reservation as to ratification/signature sans réserve de ratification
R: Ratification
A: Accession/adhésion

Country	Date of entry into force	
Algeria	04. 03. 1984	(A)
Australia	21. 12. 1975	(S)
Austria	16. 04. 1983	(A)
Belgium	04. 07. 1986	(R)
Bulgaria	24. 01. 1976	(S)
Canada	15. 05. 1981	(A)
Chile	27. 11. 1981	(A)
Denmark	02. 01. 1978	(A)
Egypt	09. 09. 1988	(A)
Finland	21. 12. 1975	(R)
France	01. 10. 1986	(R)
Gabon	30. 04. 1987	(S)
German Democratic Republic	31. 11. 1978	(A)
Germany, Federal Republic of	26. 06. 1976	(R)
Ghana	22. 06. 1988	(A)
Greece	21. 12. 1975	(A)
Hungary	11. 08. 1979	(A)
Iceland	02. 04. 1978	(A)
India	01. 02. 1982	(A)
Iran	21. 12. 1975	(R)
Ireland	15. 03. 1985	(R)
Italy	14. 04. 1977	(R)
Japan	17. 10. 1980	(A)
Jordan	10. 05. 1977	(A)
Mali	25. 09. 1987	(A)
Malta	30. 01. 1989	(A)
Mauritania	22. 02. 1983	(A)
Mexico	04. 11. 1986	(A)
Morocco	20. 10. 1980	(A)
Nepal	17. 04. 1988	(R)
Netherlands	23. 09. 1980	(R)
New Zealand	13. 12. 1976	(S)
Niger	30. 08. 1987	(S)
Norway	21. 12. 1975	(S)
Pakistan	23. 11. 1976	(R)
Poland	22. 03. 1978	(A)
Portugal	24. 03. 1981	(R)

Portugal	24. 03. 1981	(R)
Senegal	11. 11. 1977	(A)
South Africa	21. 12. 1975	(S)
Spain	04. 09. 1982	(A)
Suriname	18. 07. 1985	(A)
Sweden	21. 12. 1975	(S)
Switzerland	16. 05. 1976	(R)
Tunisia	24. 03. 1981	(A)
Uganda	04. 07. 1988	(A)
Union of Soviet Socialist Republics	11. 02. 1977	(R)
United Kingdom of Great Britain and Northern Ireland	05. 05. 1976	(R)
United States of America	18. 04. 1987	(R)
Uruguay	22. 09. 1984	(A)
Venezuela	23. 11. 1988	(A)
Vietnam	20. 01. 1989	(A)
Yugoslavia	28. 07. 1977	(A)

0. Index

The index includes subjects which are mentioned only in the notes, as well as references to the notes. If a subject which is mentioned in the text is also considered in a note elsewhere, reference is made to both the text and the note (chapter and section number).

he Avocet (Recurvirostra avosetta) is one of our most beautiful wading birds, ccuring as a breeding bird in salt meadows and small lakes in wetland areas. 56 ct. of the population breed in Ramsar sites.